**NASA
Reference
Publication
1248**

1991

Monograph on Propagation of Sound Waves in Curved Ducts

Wojciech Rostafinski
Lewis Research Center
Cleveland, Ohio

National Aeronautics and
Space Administration

Office of Management

Scientific and Technical
Information Division

Preface

This monograph presents a concise, comprehensive summary of present knowledge on the propagation of acoustic waves in bends and in systems containing bends. It covers both the characteristics and limitations of sound propagation and the information now available on the design and application of bends and elbows in various acoustical systems. A nonnegligible part of this monograph discusses various approaches adopted for handling the difficult matter of mathematically treating motion in curved boundaries.

About 30 papers have been published on the various characteristics of sound propagation in curved ducts. However, in view of the many aspects of the research, it may be difficult if not impossible to form a clear picture of the present state of knowledge without going through the entire material. And because the material is spread throughout various journals, it may be inevitable that some papers will be overlooked.

This monograph gives a general review of the subject, refers to all significant contributions, identifies the outstanding features of sound wave motion in bends, emphasizes the new and important aspects of each research paper, and compares findings. It is hoped that it will prove useful to scientists, researchers, and design engineers.

The contributions are reported as closely as possible and authors' findings and comments are carefully cited. It must be understood, of course, that only highlights and major data from reviewed papers have been used. Mathematical derivations and descriptions of experimental procedures have been omitted. To obtain additional information and often deeper analyses of the various contributions, the reader is referred to the original works.

Except for the last sections of this monograph, editorial comments are generally avoided, the idea being not to judge individual contributions, but rather to use the information as published. Finally, this monograph tries to identify the many areas still requiring substantial analytical and experimental research.

Contents

1.0 Introduction

Following the publication by Lord Rayleigh of his famous *Theory of Sound* in 1878 and except for occasional, fragmentary contributions, the subject of sound wave propagation in curved ducting remained practically untouched until the early 1970's. Before that time Krasnushkin (1945) published some developments in this field and in 1969 Grigor'yan's work, a series solution of the propagation equations, appeared, but they did not attract attention.

During the 1970's many important and revealing papers were written; all are listed in alphabetical order (by the author or authors) in the reference list. Although the subject matter needs further study, the main characteristics of sound wave motion, in the absence of a mean flow, in curved ducts and duct systems containing elbows have been established.

In their introductions several authors give engineering and scientific reasons for, as well as the philosophy behind, their studies. One of the best formulations was written by Grigor'yan (1969):

> The propagation of sound and electromagnetic waves in curved waveguides has captured the interest of many researchers. This is stimulated by the tremendous theoretical importance of the problem as a logical extension of the theory of waveguide propagation of acoustic and electromagnetic fields. Moreover, the solution of problems involving wave propagation in curved waveguides also has practical significance, because almost any waveguide system incorporates couplings of straight sections by means of curved sections....In the case of acoustical phenomena one is faced with the problem of the increase, if any, in the attenuation constant of a waveguide with absorbing walls and bending of the longitudinal axis. Of no minor interest is the problem of determining the phase velocity of sound waves in curved guides from the point of view of delay lines.

In several other papers no such introduction is offered and analysis follows a simple statement of the problem and description of the boundary conditions. However, Tam (1976) remarked that "curved ducts are an unavoidable feature of most practical duct systems. An efficient procedure to compute their acoustical characteristics should therefore be valuable to engineers and architects." Cummings (1974) observed that "such curved ducts frequently form part of elbow bends in duct or piping systems.... The engineer may well have to estimate the acoustical behavior of duct bends." Myers and Mungur (1976), Cummings (1974), and later Rostafinski (1976) reminded us that three aircraft tail engine designs utilize either a curved inlet or a curved exhaust system and that such designs may reduce noise. The S-shaped duct sections used in these engine designs were first analyzed by Baumeister (1989). Preoccupation with acoustic pollution is evident in El-Raheb's (1980) introduction and also in Fuller and Bies (1978a), who mentioned it in connection with the need to silence air-conditioning ducting. El-Raheb and Wagner (1980) called attention to the matter of noise in piping systems and said that, besides piping elements such as valves, sharp bends also are responsible for turbulence noise. Finally, let me cite from the introduction of a paper by Keefe and Benade (1983), which opened a separate and important field of study: "For at least a century and a half, makers of musical instruments have debated the effects of strongly curved portions of musical air columns." Obviously, there are both scientific and practical reasons for understanding the mechanics of acoustic wave propagation in ducts bends and in piping system elbows.

Two basic physical systems are considered by various authors. One is an infinite circular bend in a rectangular-cross-section duct, as approximated in figure 1.1(a) by a coil. At the inlet section of the coil infinitesimal and harmonic oscillation of a hypothetical piston generates sound waves. Conditions at the end section are of no consequence because the assumption of an infinite coil implies that the end section will not contribute (reflections) to the solution. The second basic physical system (fig. 1.1(b)) consists of a bend of given angle Θ connected at the inlet and end sections to straight ducts. The end of this system

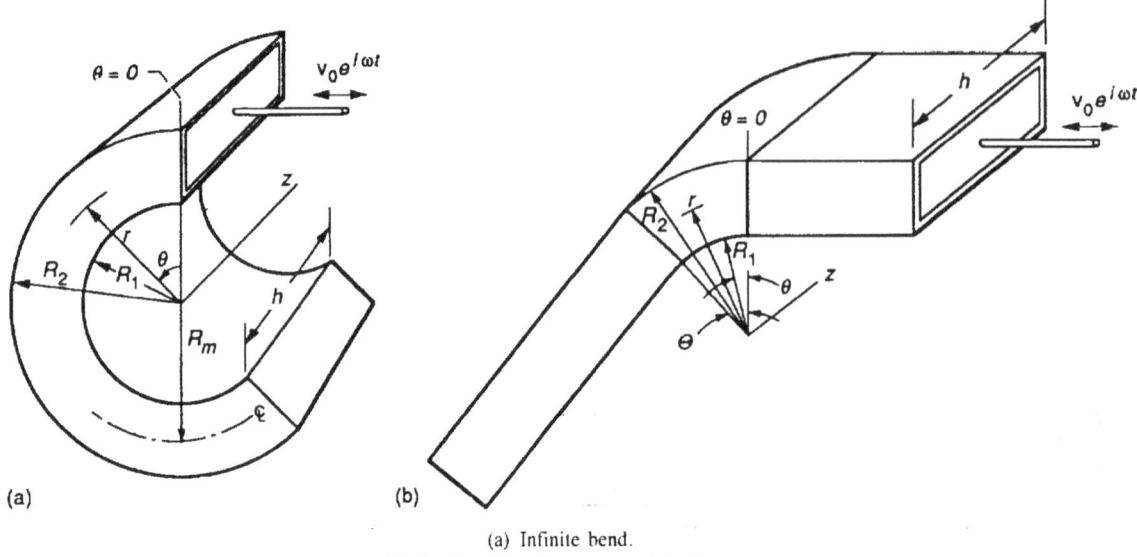

(a) Infinite bend.
(b) Bend connected to two straight ducts.

Figure 1.1.—Two physical systems most often considered.

is again at infinity so that reflected waves are not generated from the end section. The more important dimensions are indicated in both figures, and the two drawings should supplement the list of symbols in section 2.0.

Several authors treat different acoustical systems than those just described. For instance, lined ducts require additional dimensional or dimensionless symbols. Also bends formed from bent piping require different nomenclature. In all those special cases the appropriate graphs are provided as needed. For the sake of uniformity and clarity, figure and equation numbers, reference citations, and some symbols have been changed in the quotations.

2.0 More Important Symbols

a	radius ratio, R_2/R_1
c	speed of sound in free space
i	$\sqrt{-1}$
$J_\nu(kr)$, $Y_\nu(kr)$	Bessel functions of first and second kinds, of order ν and argument kr
k	wave number in free space, ω/c
k_r	radial wave number
k_z	wave number in z coordinate direction
kR_1	nondimensional acoustic wave number parameter
m,n	counting numbers, 1,2,3,...
p	acoustic pressure
R_m	bend centerline radius, $(R_2 + R_1)/2$
R_1	radius of convex (inner) wall of bend
R_2	radius of concave (outer) wall of bend
r	radius
r_2	outer radius of pipe cross section
r,θ,z	cylindrical coordinates
u	radial particle velocity
v	tangential (axial) particle velocity

$v_0\, e^{i\omega t}$	amplitude of harmonic vibration of hypothetical piston
x,y,z	rectangular coordinates
η	nondimensional wall admittance
Θ	overall bend angle
λ	wavelength
ν	angular wave number (order of Bessel functions)
ν_0	real root of basic (0,0) propagating mode
σ	susceptance
τ	conductance
ω	angular frequency, rad/sec

3.0 Early Contributions

All early (prior to 1972) papers on wave propagation in bends are fragmentary in their scope and findings. Several deal mainly with electromagnetic waves, but some contain lasting contributions to the theory and merit our attention. The most remarkable is a short, brilliant analysis in which Lord Rayleigh demonstrated, in 1878, that long waves in a curved conduit of arbitrary but infinitesimal cross section behave exactly like waves in a straight duct, the curvature having no effect on the propagation characteristics. The following is transcribed from paragraph 263 of the *Theory of Sound* (reissued in 1945):

> Hitherto we have supposed the pipe to be straight, but it will readily be anticipated that, when the cross section is small and does not vary in area, straightness is not a matter of importance. Conceive a curved axis of x running along the middle of the pipe, and let the constant section perpendicular to this axis be S. When the greatest diameter of S is very small in comparison with the wave-length of the sound, the velocity-potential ϕ becomes nearly invariable over the section; applying Green's theorem to the space bounded by the interior of the pipe and by two cross sections, we get
>
> $$\iiint \nabla^2\phi\, dV = S\cdot\Delta\left(\frac{d\phi}{dx}\right).$$
>
> Now by the general equation of motion
>
> $$\iiint \nabla^2\phi\, dV = \frac{1}{c^2}\iiint \ddot{\phi}\, dV = \frac{1}{c^2}\frac{d^2}{dt^2}\iiint \phi\, dV = \frac{S}{c^2}\frac{d^2}{dt^2}\int \phi\, dx,$$
>
> and in the limit, when the distance between the sections is made to vanish,
>
> $$\int \phi\, dx = \phi\, dx, \qquad \Delta\left(\frac{d\phi}{dx}\right) = \frac{d^2\phi}{dx^2}\, dx;$$
>
> so that
>
> $$\frac{d^2\phi}{dt^2} = c^2\frac{d^2\phi}{dx^2} \tag{3.1}$$
>
> showing that ϕ depends upon x in the same way as if the pipe were straight. By means of equation [3.1] the vibrations of air in curved pipes of uniform section may be easily investigated, and the results are the rigorous consequences of our fundamental equations (which take no account of friction), when the section is supposed to be infinitely small. In the case of thin tubes such as would be used in experiment, they suffice at any rate to give a very good representation of what actually happens.

No significant contributions to this field were made in the later years of the nineteenth century. Until about 1965, research on wave motion in curvilinear ducts remained rather rare and essentially dealt with only some aspects of the problem. Furthermore, most papers investigated the propagation of electromagnetic waves in curved ducts (waveguides). Only a few discussed the propagation of sound waves. Interestingly enough, along with analytical formulations of wave behavior in bends, there appeared a series of papers dealing with the mathematics needed to solve the equations expressed in cylindrical coordinates. This parallel effort indicates that solving the problem of bent acoustic guides required mathematical formulations (needed to handle Bessel functions of the imaginary order) then not generally available. These early contributions were discussed by Grigor'yan (1969) and Rostafinski (1970).

Two contributions to the theory of acoustic waves propagating in bends, one published by Krasnushkin (1945) and the other by Grigor'yan (1969), merit our attention. Krasnushkin approached the problem by the method of separation of variables, but in view of mathematical difficulties, he proposed a perturbation method and treated the simplified case of slightly bent tubes. Grigor'yan solved his equations by using expansion in the Taylor series, evaluated radial particle velocities, and developed an expression for the angular wave number.

4.0 Fundamental Equations

The linearized wave equation in terms of the velocity potential is

$$\nabla^2\phi = \frac{1}{c^2}\frac{\partial^2\phi}{\partial t^2}$$

with Laplacian

$$\nabla^2 = \frac{\partial^2}{\partial x^2} + \frac{\partial^2}{\partial y^2} + \frac{\partial^2}{\partial z^2} \qquad \text{in rectangular coordinates}$$

$$\nabla^2 = \frac{\partial^2}{\partial r^2} + \frac{1}{r}\frac{\partial}{\partial r} + \frac{1}{r^2}\frac{\partial^2}{\partial\theta^2} + \frac{\partial^2}{\partial z^2} \qquad \text{in cylindrical coordinates}$$

where the velocity potential ϕ is a function of three dimensions and of time.

The particle velocities are given by

$$\frac{1}{r}\frac{\partial\phi}{\partial\theta} = v(r,\theta,z,t), \text{ the tangential component} \qquad (4.1)$$

$$\frac{\partial\phi}{\partial r} = u(r,\theta,z,t), \text{ the radial component} \qquad (4.2)$$

$$\frac{\partial\phi}{\partial z} = w(r,\theta,z,t), \text{ the component in the direction of the two flat walls of the duct} \qquad (4.3)$$

The acoustic pressure is given by

$$p(r,\theta,z,t) = -\rho\frac{\partial\phi}{\partial t} \qquad (4.4)$$

From this point on, analytical treatment becomes a boundary value problem and, generally speaking, three methods of solving the problem have been used. One, classical but severely limited by mathematical difficulties, relies on the separation of variables. The second, of limited applicability, relies on perturbation principles. The third, of limited conceptual value but of immense power to yield information, consists of various numerical computerized methods.

The most general, three-dimensional formulation of the problem of sound propagation in a bent duct was outlined by several authors but solved and applied only by Ko (1979), for an acoustically lined duct, and by El-Raheb (1980), for a multiplane duct system with four bends. Because the important characteristics of wave propagation in bends depend on the curvature of the duct and not on its height, and as indicated by Fuller and Bies (1978a), since there is no discontinuity in duct height, it is satisfactory to use a two-dimensional coordinate system. Consequently, analyses of bent-duct acoustics by most authors treat only the two-dimensional case. Newer contributions to the theory of sound propagation in bends treat special cases of nonrectangular-cross-section bent piping. Besides a short contribution by Prikhod'ko and Tyutekin (1982), there is one (of limited scope) by Ting and Miksis (1983), a study on bent elliptical piping by Furnell and Bies (1989), and an extensive study of 90° bent piping by Firth and Fahy (1984).

Several papers (which are cited in subsequent sections) treat two segments of straight duct connected by a bend and, as mentioned before (El-Raheb, 1980), multibend ducting. These cases call for formulating wave propagation in straight ducts, a known and proven technique, but more importantly for adjusting the fields at the junctions in straight-bent ducts. The compatibility relations based on continuity of tangential particle velocities and pressures or on continuity of both tangential and radial particle velocities at junctions allow us to determine (1) constants of integration, (2) propagating modes, and (3) amplitudes of the nonpropagating modes (attenuated, evanescent waves traveling both upstream and downstream from each junction).

Solving the equations given so far may call for evaluating eigenvalues, it may require integration, it may be based on direct numerical evaluation of the differential equations, or it may require separation of variables. The last method is important because it allows a good physical interpretation of the characteristics of motion in bends. The main equations of this method are given here.

Since the linearized two-dimensional wave equation in cylindrical coordinates is known to be separable, the Helmholtz equation may be broken up into a set of ordinary differential equations, each including a separation constant. In the important two-dimensional case the solution may be assumed to be

$$\phi = \Re(r)\ \Phi(\theta)\ \Pi(t)$$

with

$$\frac{1}{c^2}\frac{\Pi''}{\Pi} = -k^2$$

and

$$\frac{\Phi''}{\Phi} = -\nu^2$$

which leads to

$$\Pi = e^{i(\omega t + \alpha)}$$

and

$$\Phi = a_\nu \cos \nu\theta + b_\nu \sin \nu\theta \qquad \nu \neq 0$$

$$\Phi = d\theta + g \qquad \nu = 0$$

Finally

$$\Re'' + \frac{1}{r}\Re' + \left(k^2 - \frac{\nu^2}{r^2}\right)\Re = 0 \qquad (4.5)$$

where

$$\Re = A_\nu J_\nu(kr) + B_\nu Y_\nu(kr) \qquad (4.6)$$

which is the characteristic function of the problem. Here $J_\nu(kr)$ and $Y_\nu(kr)$ are the Bessel functions of the first and second kinds. The argument of these functions should be written $(k_r r)$, but in a two-dimensional case it may be simplified to (kr). Consider the three-dimensional case where $k_r^2 = k^2 - k_z^2$. In a two-dimensional case with $(k_z = 0)$ it simplifies to $k_r = k = \omega/c$, as used for convenience in the characteristic equation.

Since superposition of solutions is allowed, the general solution may be written in mathematical physics notation in the form

$$\phi = \sum_{\nu \in C} e^{i(\omega t + \alpha)}(a_\nu \cos \nu\theta + \sin \nu\theta)\left[A_\nu J_\nu(kr) + B_\nu Y_\nu(kr)\right]$$

where α is a possible phase lag and the solution $\nu = 0$ (linear dependence) is discarded and where C is a finite set of points in a complex plane to be determined in order to satisfy the boundary conditions. Writing in a more familiar form, a double summation can be used. Summation of all wave eigenmodes traveling in the positive and negative directions in a bend gives

$$\phi = \sum_{m=0}^{\infty} \sum_{n=0}^{\infty} A_{mn}\Re_{mn}(r,\theta)e^{i(\omega t - \nu_{mn}\theta)} \qquad (4.7)$$

where $m,n = 1,2,3,\ldots$ and all coefficients A_{mn} are complex amplitudes.

In order to satisfy the partial differential equation and the boundary conditions for acoustically lined (or perfectly rigid) cylindrical walls, a characteristic equation is established whose roots ν's are the characteristic values (eigenvalues) of the problem that yield nontrivial solutions. Grigor'yan (1969) formulated a two-dimensional equation for different types of lining on the two curved walls; later Ko (1979) wrote a complete three-dimensional equation for lined ducts. Simpler and sufficient for general discussion is the two-dimensional eigenvalue equation for identical linings on the two curved walls as given by Rostafinski (1982):

$$[J_\nu'(kR_1)Y_\nu'(kR_2) - J_\nu'(kR_2)Y_\nu'(kR_1)] + i\eta[J_\nu'(kR_1)Y_\nu(kR_2)$$

$$- J_\nu(kR_2)Y_\nu'(kR_1) - J_\nu(kR_1)Y_\nu'(kR_2) + J_\nu'(kR_2)Y_\nu(kR_1)]$$

$$+ \eta^2[J_\nu(kR_1)Y_\nu(kR_2) - J_\nu(kR_2)Y_\nu(kR_1)] = 0 \qquad [4.8]$$

where $\eta = -\rho c u/p$ (i.e., a dimensionless wall admittance u/p). In general, $\eta = \tau + i\sigma$ is complex, τ is the conductance of the walls, and σ the susceptance of the walls. It is clear that using $\eta = 0$ in Eq. [4.8] leaves only the expression of the cross products of the derivatives of the two Bessel functions, which corresponds to the case of the hard-walled bend.

For this two-dimensional case, as explained before, k_r is written as k for convenience, in the arguments of the Bessel functions.

Various analytical techniques used by different authors to solve this equation are described, and their approaches discussed, in the next section.

5.0 Analytical Solutions

Just three basic mathematical methods have been used to calculate the characteristics of sound propagation in bends: perturbation, separation of variables, and direct numerical techniques. The method of perturbation, powerful but often limited, has been used by physicists for studying electromagnetic waves in cylindrically curved waveguides. By the nature of this method, only the limiting case of propagation in slightly bent tubes could be considered. Results of such analyses cannot be extrapolated to situations involving high-curvature bends. Of the 20 authors who did contribute new developments to the subject of acoustics in bends, only Krasnushkin (1945) and to a degree Ting and Miksis (1983) and Prikhod'ko and Tyutekin (1982) used the perturbation method.

Fifteen authors relied on the method of separation of variables and subsequently proceeded, to a greater or lesser degree, with numerical integration and other numerical techniques, to obtain data that allow the formulation of laws for sound propagation in bends. Only Tam (1976), El-Raheb and Wagner (1980), and Baumeister (1989) used direct numerical techniques such as the Galerkin method, Green functions, and iterative methods. Also Cabelli and Shepherd (1981) relied on direct finite element computation, but they evaluated the acoustical properties of a rounded 90° corner rather than a 90° bend. Furnell and Bies (1989) relied, to a degree, on the calculus of variations. All techniques used by the various authors are tabulated, in alphabetical order, in table 5.1. Rostafinski (1972) used the traditional and most used method of separation

TABLE 5.1.—MATHEMATICAL METHODS OF ANALYSIS

Baumeister and Rice (1975)	Finite difference formulation.
Baumeister (1989)	Finite-element Galerkin formulation of the wave equation and of the boundary conditions in dimensionless form.
Cabelli (1980)	Numerical solution (finite difference equations) of the two-dimensional Helmholtz equation and boundary conditions. Experiments.
Cabelli and Shepherd (1981)	Finite element technique. The subject is not exactly a cylindrical bend—more a rounded corner. Experiments.
Cummings (1974)	Separation of variables followed by numerical integration using Simpson's rule and solution of simultaneous equations by Crout's method. Experiments.
El-Raheb (1980)	After separation of variables, eigenfunction expansion of the Helmholtz equation. Matrix of coefficients of compatibility equations solved by inversion and factorization.
El-Raheb and Wagner (1980)	Direct solution of the finite difference equation by using square orthogonal finite difference grid. Also Green's function formulation. Good agreement between the two.
Firth and Fahy (1984)	Solution of the Helmholtz equation in toroidal coordinates by using approximate formulations for the radial functions in the pipe bend.
Fuller and Bies (1978a,b)	Separation of variables followed by power series expansion of the Bessel and Neumann functions, computer integrations by Simpson's rule, and solving the matrix by Crout's method. Experiments.
Furnell and Bies (1989)	Procedure based on the Rayleigh-Ritz method for obtaining numerical approximations of the acoustic modes. Matrices characterize modal transmissions.
Grigor'yan (1969 and 1970)	Separation of variables followed by formulation of expansion in Taylor series. Numerical data with computer's help.

TABLE 5.1—Concluded.

Keefe and Benade (1983)	Separation of variables, transmission line model for impedance, and integrations by Simpson's rule. Experiments.
Ko and Ho (1977)	Separation of variables and eigenfunctions solved numerically by iterative process.
Ko (1979)	Separation of variables, numerical evaluation of eigenvalue equation, and successive approximations by using Newton-Raphson method.
Krasnushkin (1945)	Separation of variables and perturbation method.
Myers and Mungur (1976)	Separation of variables, numerically (fourth-order Runge-Kutta scheme integrations) obtained eigenvalues and amplitudes.
Osborne (1974)	Evaluation of eigenvalues by using computer code for iteration of interpolations.
Osborne (1976)	Separation of variables and numerical integrations by multiple-strip Simpson's rule and phase angles obtained from vector diagrams. Experiments.
Prikhod'ko and Tyutekin (1982)	Helmholtz operator and perturbation.
Rostafinski (1970)	Separation of variables and one integral obtained by Simpson's rule.
Rostafinski (1972)	Separation of variables and one integral obtained by Simpson's rule.
Rostafinski (1974a,b; 1976)	Separation of variables and numerical iterations.
Rostafinski (1982)	Separation of variables and numerical interpolations and iterations.
Tam (1976)	Galerkin's method with iterative algorithm to solve the eigenvalue matrix.
Ting and Miksis (1983)	Perturbation method for bend in tubular region of arbitrary shape.

of variables in his work on the propagation of long waves in bends of arbitrary sharpness. Because it is the only paper with explicit, not numerical, formulation of the eigenvalues (angular wave numbers) of the propagating and evanescent modes, the more interesting parts of this analysis are given in appendix A.

On the other hand, studies that used separation of variables and numerically solved the characteristic equation to obtain both the eigenvalues and the compatibility equations provide more useful data. Because of mathematical difficulties in formally evaluating Bessel functions of complex orders, all approaches must to a degree use numerical calculations. These studies, aided by modern numerical techniques, give insight into acoustic propagation variables. There is no question now that direct computational methods are effective and yield valuable information.

6.0 Angular Wave Number

Analytical solutions based on the method of separation of variables lead to formulation of a characteristic equation and proceed to determine the roots of this equation, the eigenvalues. These roots are orders of the Bessel functions of the first and second kinds that form the equation. It is worthwhile to bring up an observation made by Ko and Ho (1977), who state that the eigenvalue equation for a cylindrical bend in a hard-wall duct is identical to that for a straight, annular hard-wall duct. However, a basic difference exists between these two cases. The eigenvalue for a cylindrical bend is the order itself of the Bessel functions of the two kinds, for a given argument, when the acoustic wave propagates around the bend. In contrast, in the better-known case, the eigenvalue for a straight annular duct is the argument of the Bessel functions of the first and second kinds, for a given order, when the acoustic wave propagates along the axial direction. The eigenvalues of the characteristic equation, orders of the Bessel functions, were named by Krasnushkin (1945) "angular wave numbers," and this name has been adopted universally. Since the order of the Bessel functions is indicated by the letter v, the angular wave number almost always is known by this symbol.

The angular wave number appears in the wave equation exponential term $e^{i(\omega t - v\theta)}$ and is a nondimensional number as opposed to the wave number k, a proportionality constant in the expression $e^{k(ct-x)}$ in the theory for waves propagating in the rectilinear x direction; the dimension of k is, of course, length to the -1 power. Angular wave numbers v may be real fractional numbers (and zero at the cutoff), pure imaginary numbers, or complex numbers. Because v are orders of the Bessel functions, their evaluation has been for decades the stumbling block in mathematical formulations of wave motion in bends.

6.1 Hard-Wall Bends

The characteristic equation for hard-wall ducts is greatly simplified and becomes treatable in situations where series expansions of the Bessel functions are possible. In general, however, numerical techniques are necessary and yield good answers. The first successful attempt to calculate v was done by Rostafinski (1970 and 1972). Because his study was limited to long waves and extremely low frequencies, he could limit series expansions to two or three terms and get results without recourse to numerical methods. The procedure is a classical analysis in which the infinite set of pure imaginary $v_m = i(m\pi/(\ln a))$, $m = 1,2,3,...$ is obtained and next the single real root v_0 of the basic (0,0) propagating mode is calculated by the perturbation method. Details of this analysis are given in appendix A.

Rostafinski's expression for the angular wave number shows that it is a function of nondimensional frequency as given by the parameter kR_1 (the wave number in free space times the inner radius of the cylindrical bend) and of the bend radius ratio $a = R_2/R_1$:

$$v_0 = \left(\frac{2\dfrac{a^2 - 1}{\ln a}}{4(kR_1)^{-2} + 1 + a^2 + \dfrac{a^2 - 1}{\ln a}} \right)^{1/2} \tag{6.1}$$

For $a \to 1$ with $\lim_{a \to 1} \dfrac{a^2 - 1}{\ln a} = 2$,

$$v_0 \to \left[\frac{1}{(kR_1)^{-2} + 1} \right]^{1/2} \approx kR_1$$

Using expansion in the Taylor series, Grigor'yan (1969) obtained

$$v_0 = a(kR_1)\left(\frac{3a - 1}{5a - 3} \right)^{1/2}$$

This equation tends to $v = kR_1$ for $a \to 1$ as required, but when $a \gg 1$, its values become unreliable. Calculations indicate that Rostafinski's expression, even for a of 5 to 10 (sharp bends), yields v_0's that verify the differential equation, but Grigor'yan's does not. Obviously, expansions in the Taylor series yield only approximate values.

Cummings (1974) did not calculate the angular wave number but gave an interesting discussion of it. On the other hand, Osborne (1968) restricted ν_0 to certain values (simply for convenience) by assuming that the pressures at two angular positions in the duct are the same; this restriction was unrealistic in practice for propagating waves, and ν_0 must be allowed to take on whatever value is determined by the boundary conditions at the walls.

Cummings (1974) defined the cutoff frequencies of the modes in a curved section. This condition of $\nu_0 = 0$ in the case of $k_z = 0$ corresponds to cutoff frequencies for the radial modes in a straight annular duct. Osborne (1976) published an interesting table of roots $k_{r_m}R_1$, given here as table 6.1, when the radial wave number is

$$k_{r_m} = \left[\left(\frac{\omega}{c} \right)^2 - \left(\frac{m\pi}{h} \right)^2 \right]^{1/2}$$

The subscript m indicates the propagation mode taken into consideration and h is the depth of the duct in the z coordinate direction.

The first evaluation of the propagation constants for the evanescent waves was done by Rostafinski (1972), who calculated the angular wave numbers for long waves. His equation for the cross products of the derivatives of the Bessel functions of the first and second kinds was simplified by using the relation

$$Y'_\nu(kr) = \frac{(\cos \pi\nu)J'_\nu(kr) - J'_{-\nu}(kr)}{\sin \pi\nu}$$

and thus the characteristic equation for a hard-wall duct containing only Bessel functions of the first kind of positive and negative order becomes

$$F_{\nu_m}(r) = J'_{\nu_m}(kR_1)J_{-\nu_m}(kr) - J_{\nu_m}(kr)J'_{-\nu_m}(kR_1) \tag{6.2}$$

When the condition of zero radial vibration at $r = R_2$ is applied, this equation yields

$$(\sin \pi\nu)^{-1}\left[J'_{\nu_m}(kR_1)J'_{-\nu_m}(akR_1) - J'_{\nu_m}(akR_1)J'_{-\nu_m}(kR_1) \right] = 0 \tag{6.3}$$

Equation (6.3) is equation (3) in Rostafinski (1972). Calculation of the imaginary roots (eigenvalues of the evanescent waves) was straightforward. Because the only existing tables of the Bessel functions of the imaginary order, by Buckens (1963), were restricted to imaginary integer numbers, the radius ratio

TABLE 6.1—ROOTS $k_{rm}R_1$ of $J'_0(k_{rm}R_1)Y'_0(k_{rm}R_2) - J'_0(k_{rm}R_2)Y'_0(k_{rm}R_1) = 0$

[From Osborne (1976).]

Radius ratio, $a = R_2/R_1$	Mode					
	1	2	3	4	5	6
1.2	15.728	31.426	47.131	62.837	78.544	94.251
1.5	6.322	12.586	18.863	24.143	31.424	37.706
2.0	3.197	6.312	9.445	12.581	15.720	18.860
2.5	2.157	4.223	6.307	8.395	10.486	12.576
3.0	1.636	3.179	4.738	6.303	7.870	9.441
4.0	1.112	2.134	3.170	4.210	5.253	6.298
∞	3.832	7.016	10.174	13.324	16.471	19.616

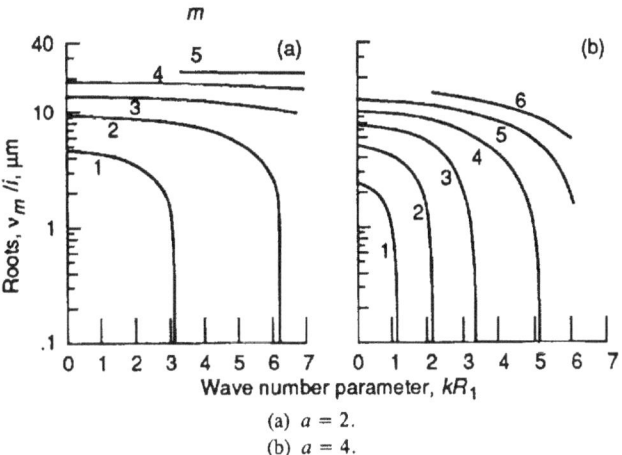

(a) $a = 2$.

(b) $a = 4$.

Figure 6.1.—First five roots for ducts with radius ratios a of 2 and 4 in range of wave number parameter. From Rostafinski (1976).

$a = R_2/R_1 = 1.8744$ was used; it gave

$$v_m = \frac{im\pi}{\ln a} = 5im \qquad m = 1,2,3,\ldots$$

Error in the calculated (on an electrical office calculator) cross product of the derivatives of the Bessel functions of the imaginary order did not exceed 0.12 percent. Data on angular wave numbers for the evanescent modes for ducts with a of 2 and 4 are given by Rostafinski (1976). Calculation was done by using Bessel functions of order $(n + 1/2)$, $n = 0,1,2,3,\ldots$. The two graphs in figure 6.1 clearly indicate how the roots decrease with increasing frequency until cutoff values are reached when $v_0 = 0$ (not shown). Rostafinski's values of kR_1 when $v_0 = 0$ are approximate; those given in table 6.1 are exact.

El-Raheb (1980) also used $a = 1.8744$ to calculate, for various parameters, an extensive set of the angular wave numbers of the basic mode and several higher modes as a function of a nondimensional acoustic wave number parameter kR_1. Although restricted to a single radius ratio, his graphs illustrate well the nature of changes and the ability of the basic acoustic mode to propagate in curved ducts (fig. 6.2).

Both Rostafinski (1974a) and Ko and Ho (1977) calculated a number of angular wave numbers for various propagation modes. To profit from closed-form solutions for Bessel functions of order $(n + 1/2)$, $n = 0,1,2,3,\ldots$, Rostafinski used an inverted method to determine the characteristics of motion. With this method the characteristic equation could be written in a simple closed form (NBS, 1964) and was solved by iterating for the arguments kR_1 that would satisfy the boundary conditions. The same technique allowed him to construct, for comparison, the wave numbers for a straight duct of the same sidewall spacing. The results are given in figure 6.3. Similarly, Ko and Ho (1977) produced a set of curves v_0 (see fig. 6.4) by using numerical techniques and nondimensional frequency $f_\ast = f \cdot 2R_1/c$, where f is the frequency and c is the speed of sound. The curves pertain to bends with a of 1.11, 1.25, and 1.43 ($1/a = 0.9, 0.8, 0.7$). Later Fuller and Bies (1978b), using an iterative process on a computer, solved for the angular wave numbers (roots) of the characteristic equation, consisting of the derivatives of the Bessel and Neumann functions, and showed the results for two radius ratios on a graph (fig. 6.5). Figures 6.3 to 6.5 can be used to interpolate, with satisfactory accuracy, values of angular wave numbers for any particular engineering application. All eigenvalues calculated by various methods and for a quite wide range of parameters constitute nevertheless a coherent and useful system of data.

An important contribution to the process of calculating the angular wave number has been made by Tam (1976). He used a method (a computer program) that contains an iterative algorithm that uses the Galerkin method to solve the matrix eigenvalue problem. In this way he effectively avoided the eigenvalue equation, which involves (as we well know by now) complicated Bessel functions of noninteger order. Besides figure 6.6, which gives calculated eigenvalues of the propagating modes in a square-cross-section duct for $R_1 = 3.0$ and mode 2, Tam gives extensive tables of eigenvalues for selected (example) parameters. His method was

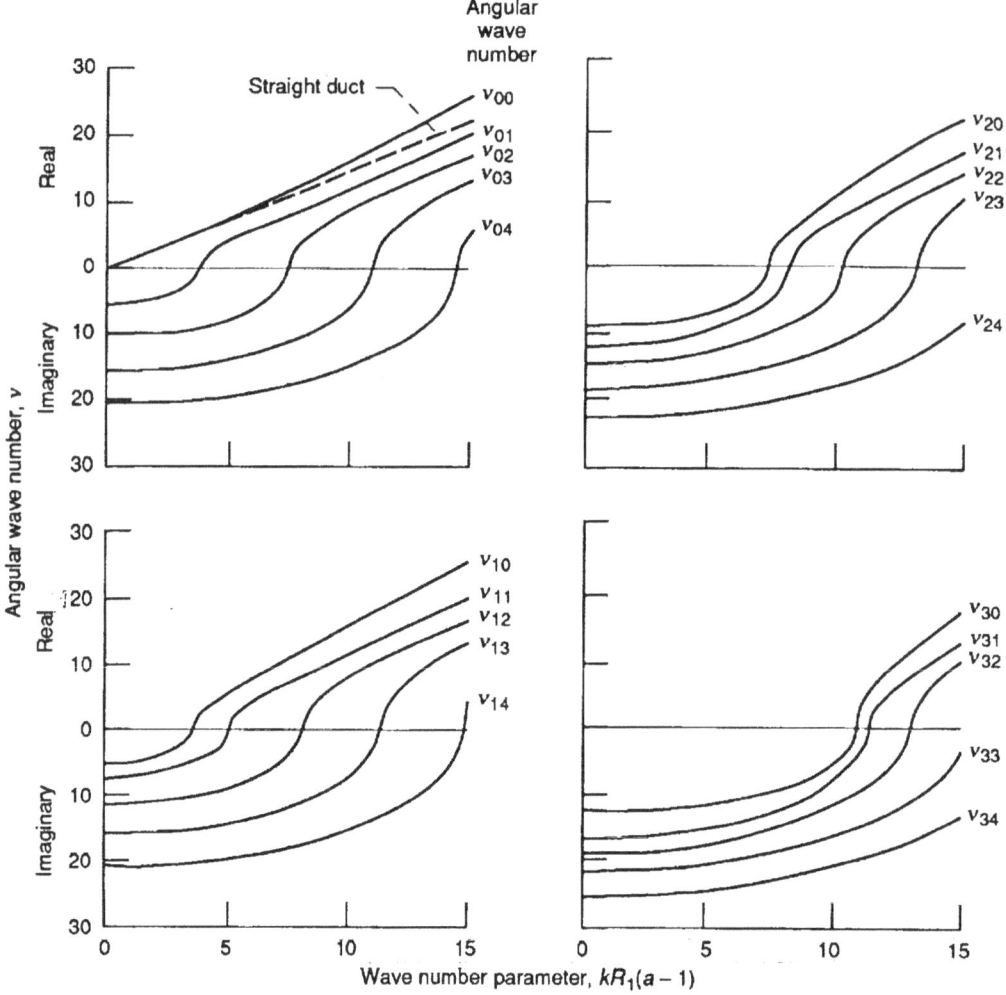

Figure 6.2.—Angular wave number for several acoustic modes as function of wave number parameter. Radius ratio, $a = 1.8744$. Note the cutoff frequencies at $\gamma = 0$. From El-Raheb (1980).

developed primarily for rectangular ducts with rigid walls. With some modifications this procedure should be available for soft-wall ducts.

6.2 Acoustically Lined Bends

The eigenvalues (the angular wave numbers corresponding to propagation of waves in acoustically lined curved ducts) are complex, as are, of course, the wave numbers corresponding to propagation in lined straight ducts. The basic difficulty in solving equations of motion in lined bends resides in our inability to split the final complex equation containing exponential and trigonometric functions into real and imaginary parts. Were this possible, it still would not guarantee direct answers, but it would greatly help because a single term would be sufficient to satisfy the characteristic equation.

With complex expressions two real numbers $\nu = \nu_1 + i\nu_2$ must be found, so that one is forced to rely on the sometimes tedious method of successive approximations. Rostafinski (1982), in studying long waves at extremely low frequencies, used series expansion of Bessel functions of which he could retain only one or two terms. With nondimensional wall admittance $\eta = -\rho c u / p = \tau + i\sigma$ (where τ is the lining conductance and σ is the lining susceptance), solving the eigenfunction by expanding in series and retaining only the first terms of the series, rearranging, and eliminating small terms of higher order gave

$$\tau^2 - \sigma^2 + 2i\tau\sigma + (\sigma - i\tau)\frac{a+1}{akR_1}\frac{a^\nu + a^{-\nu}}{a^\nu - a^{-\nu}} - \frac{\nu^2}{a(kR_1)^2} = 0. \qquad [6.4]$$

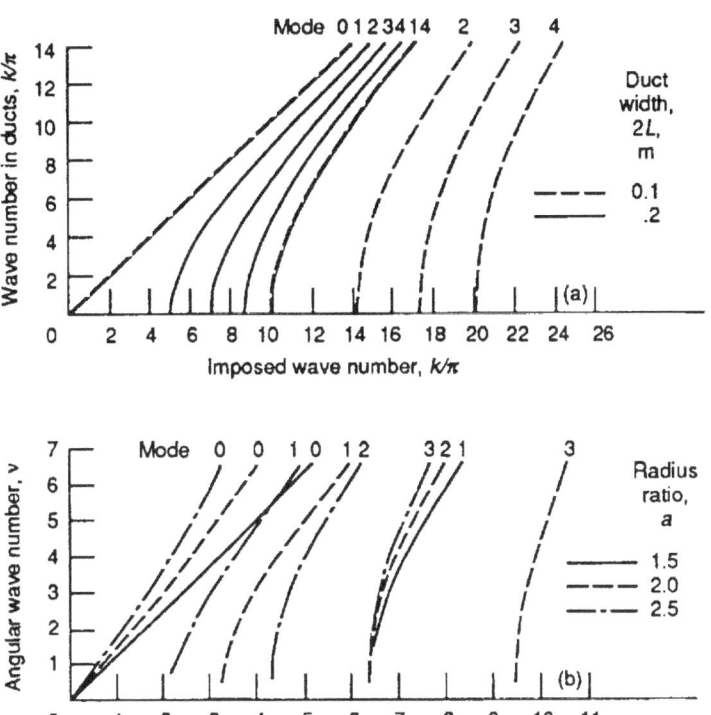

(a) In two straight ducts of different widths.
(b) In three bends of different radius ratios.

Figure 6.3.—Characteristics of motion in bends and ducts. From Rostafinski (1974).

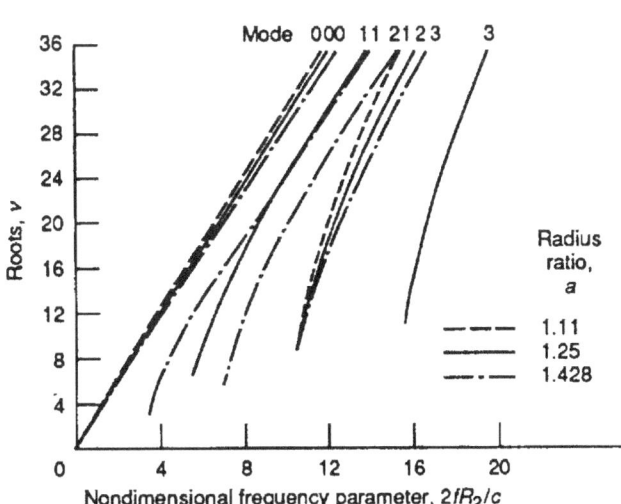

Figure 6.4.—Eigenvalues (roots) for rigid curved duct with three bends. From Ko and Ho (1977).

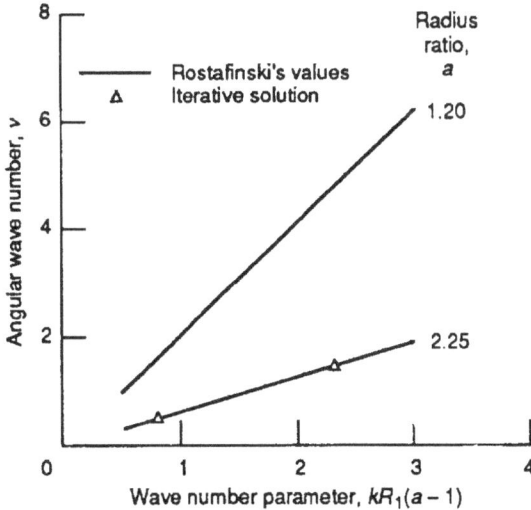

Figure 6.5.—Solutions of curved duct characteristic equation for two radius ratios *a*. From Fuller and Bies (1978b).

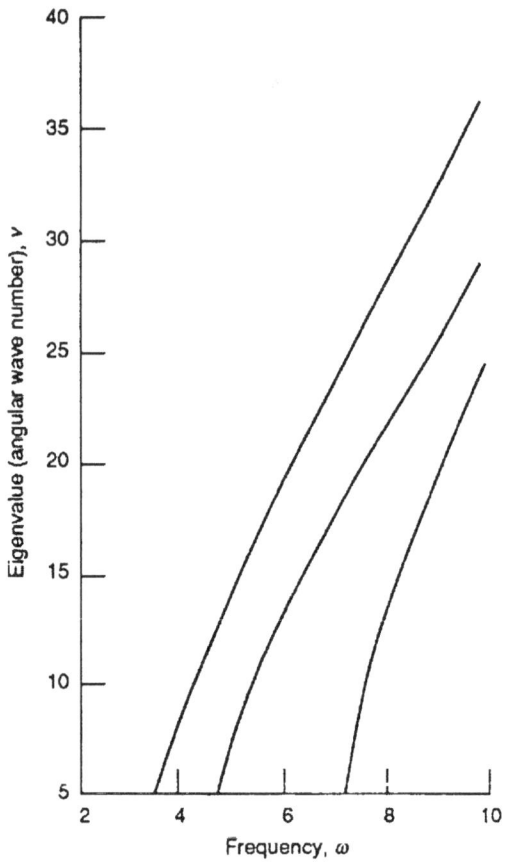

Figure 6.6.—Eigenvalues of propagating modes as function of frequency for square-cross-section duct. Radius of convex (inner) wall of bend, $R_1 = 3$. From Tam (1976).

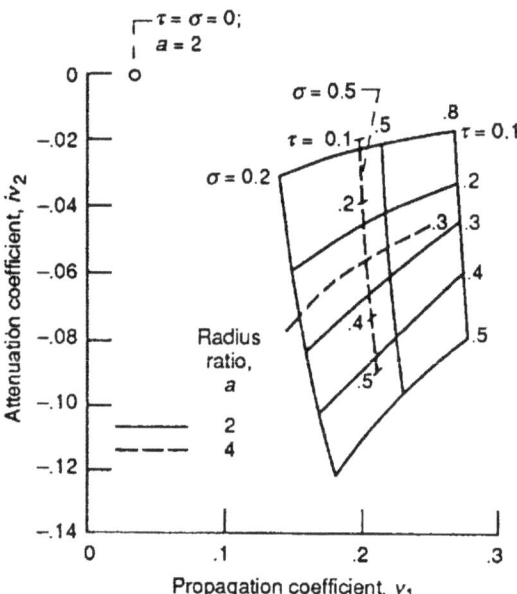

Figure 6.7.—Real and imaginary parts of complex angular wave numbers for curved duct. Admittance, $\eta = \tau + i\sigma$; angular wave number, $\nu = \nu_1 + i\nu_2$. From Rostafinski (1982).

TABLE 6.2—COMPLEX ANGULAR WAVE NUMBERS FOR CURVED LINED DUCTS FOR
$kR_1 = 0.02$ AND RANGE OF LINING PARAMETERS

Susceptance, σ	Conductance, τ	Radius ratio, a			
		2		4	
		Propagation, ν_1	Attenuation, ν_2	Propagation, ν_1	Attenuation, ν_2
0.2	0.1	0.13546	−0.03201	--------	--------
	.2	.14460	−.05997	--------	--------
	.3	.15577	−.08351	0.14225	−0.07680
	.4	.16735	−.10364	--------	--------
	.5	.17873	−.12131	--------	--------
0.5	0.1	0.20931	−0.02077	0.19199	−0.01925
	.2	.21224	−.04097	.19462	−.03797
	.3	.21672	−.06019	.19864	−.05580
	.4	.22234	−.07823	.20367	−.07256
	.5	.22873	−.09505	.20938	−.08823
0.8	0.1	0.26415	−0.01651	--------	--------
	.2	.26565	−.03283	--------	--------
	.3	.26805	−.04880	0.24649	−0.04564
	.4	.27125	−.06480	--------	--------
	.5	.27511	−.07925	--------	--------

For $\tau = \sigma = 0$, the conditions for stiff walls, Eq. [6.4] yields $a^\nu = a^{-\nu}$ or $a^{2\nu} = 1$, and $\nu = in\pi/\ln a$, pure imaginary roots derived before in [Rostafinski (1972)]. Equation [6.4] also allows verification that for finite values of τ and σ, there are no pure imaginary or simple real roots. Hence, ν must be complex. Roots, which are complex angular wave numbers, have been obtained by successive approximations. It helped to know that both the real and the imaginary parts of ν must be small because the real root for the case of the hard walls is small. In applying Eq. [6.4], calculations have been made for the following set of parameters: $k = 0.1$; $R_1 = 0.2$; $a = 2$ and 4; $\tau = 0.1$, $0.2,\ldots,0.5$; and $\sigma = 0.2$, 0.3, 0.5, and 0.8. The calculated complex roots of Eq. [6.4], which are also complex propagation constants, are listed in Table [6.2] and are shown in Figure [6.7]. It will be noted that quantities related to attenuation are one order of magnitude smaller than terms related to propagation. Attenuation increases with increasing τ, but attenuation is less pronounced with a wider duct. For small values of conductance τ, susceptance σ does not greatly influence attenuation; with higher τ, however, large changes occur with changing σ. The percentage change in attenuation at constant τ and varying σ is nearly twice as great for the lower range of τ as for the largest τ studied. On the same graph, the angular wave number for the rigid, curved wall case (i.e., for $\tau = \sigma = 0$, and $a = 2$) is also shown. Presence of acoustical lining on curved walls markedly increases the values of the propagation constants ν_1.

The three-dimensional analysis of lined bent ducts described by Ko (1979) is extensive and offers detailed tables of eigenvalues (the angular wave numbers for selected frequencies, bend geometries, and lining characteristics). Using those values, he calculated sound attenuation in bends. His curves are given in figure 11.10.

Myers and Mungur (1976) stated that ν being complex renders the formal expression for the characteristic function of limited practical value in determining the radial functions. Nevertheless, they calculated those wave numbers numerically and determined normalized pressure modes for a rigid-wall bend for given frequency and bend dimensions and for a given wall admittance.

7.0 Phase Velocity

Evaluation of the phase velocity in curved ducts is one of the major and more important areas of analysis done by several authors. Phase velocity happens to be an important wave characteristic in engineering applications of bends.

Cummings (1974) established that, for most engineering purposes, the curved sections of ducts can be regarded as equivalent to a straight duct but not of the same median length. In fact, Rostafinski (1970) has shown that the phase velocity of long waves is higher when averaged across a bend cross section than in straight ducts. In bends the phase velocity is $\dot\theta = \omega/\nu_0$, as obtained from $e^{(\omega t - \nu_0 \theta)}$; in straight ducts it is $\dot x = \omega/k = c$, as given by $e^{(\omega t - kx)}$. To compare the two velocities, average the tangential phase velocity $\dot\theta r = \dot s$ over the duct width (from R_2 to R_1) and obtain $\dot s = R_m\omega/\nu_0$. The ratio of the two velocities is $\dot s/\dot x = kR_m/\nu_0$ and equals $kR_1(1 + a)/2\nu_0$, as written by El-Raheb (1980). As shown in figure 7.1 (Rostafinski, 1970) the phase velocity of long waves is higher in bends than in straight ducts and increases with the sharpness of the bend. Note that in the limit when $a \to 1$ the phase velocity of a slightly bent duct equals the phase velocity in its straight segment. This last point has been noted by Grigor'yan (1969), who stated that the fundamental mode in a slightly bent duct propagates along a median circle with the wave number for the straight duct. As indicated in the section Early Contributions, Lord Rayleigh arrived at this conclusion first and thus set a limit which the equation of wave motion in curved ducts must satisfy.

Cummings (1974) calculated the phase velocities in bends for a range of frequencies. He discovered that at $kR_1(a - 1)/\pi = 0.5$, which is the wave number of the imposed acoustic disturbance multiplied by the bend width over π equals 0.5, the phase velocities in bends of all sharpnesses equal that in a straight duct but that at frequencies higher than 0.5 they are always lower. This is illustrated in figure 7.2, adapted from Cummings (1974).

El-Raheb (1980) expanded on this subject and warned that little information may be obtained from calculating phase velocity beyond one-half of the first cutoff wave number (at $kR_1(a - 1)/\pi = 0.5$) because then coupling between modes becomes stronger and the phase velocity may depend on the relative magnitude of the phases. These considerations were further developed by El-Raheb and Wagner (1980). They indicated that in the expression for phase velocity ratio $\dot s/\dot x = kR_1(a + 1)/2\nu_0$ for frequencies below the first cutoff

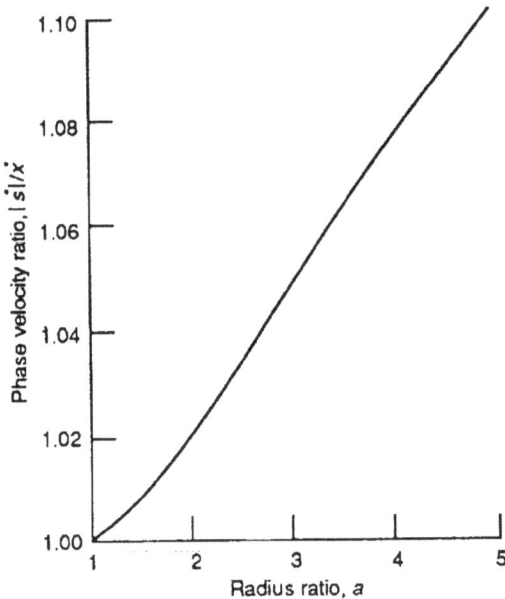

Figure 7.1.—Propagation of long waves in curved ducts. Comparison of phase velocity in bends with phase velocity in straight ducts. From Rostafinski (1970).

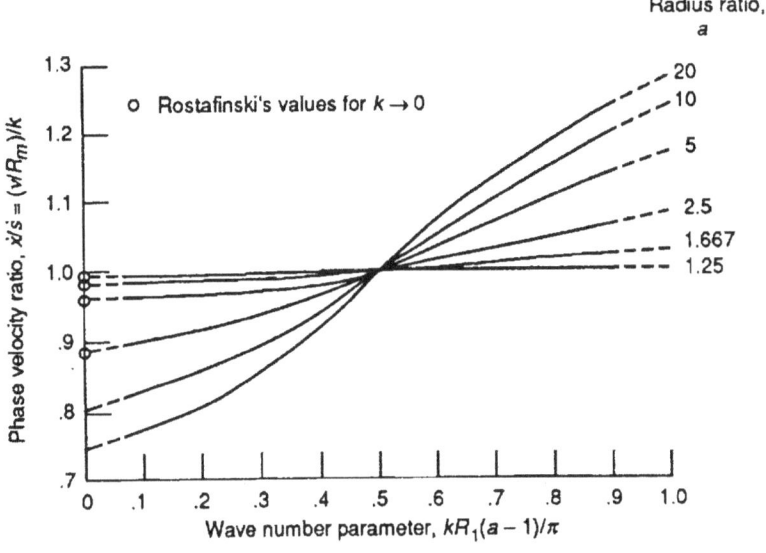

Figure 7.2.—Design chart for rectangular section bends. From Cummings (1974).

and for $a < 3$, ν_0 varies linearly with a, but that at frequencies approaching cutoff from below this variance becomes weakly quadratic with increasing rate. For bends with $a > 3$ the phase velocity ratio becomes less than unity at smaller values of kR_1, meaning that the phase velocity reversal is achieved at frequencies lower than 0.5.

As an interesting footnote, recall that Buchholz (1939) in his work on electromagnetic waves in curved waveguides stated that waves are propagated more rapidly in (slightly) bent tubes than in straight tubes. Krasnushkin (1945) commented that Buchholz's statement is a misunderstanding, since no matter how much a tube is bent there is a straight line (normal to the radii) on each wave front where the velocity is equal to the velocity in straight segments. More precisely, in any bend a straight line can be found that divides the wave front into two parts. For the external part of the front the phase velocity is greater and for the internal part it is less than the wave velocity of the same number for a straight tube.

The phenomenon of phase velocity in bends being a function of frequency and bend sharpness has been experimentally proven by Fuller and Bies (1978a,b) and applied in designing a reactive acoustic attenuator, which consists of a bent duct with a curved axial partition. Their two interesting papers and findings are discussed in section 13.0.

8.0 Particle Velocity and Pressure Distribution

The pressure distribution in bends has been investigated by many researchers. Grigor'yan (1969) and Rostafinski (1972) first published the results of their calculations of particle velocity distributions in bends. Grigor'yan showed distribution of radial velocities—both propagating and evanescent. Rostafinski evaluated the tangential particle velocities (sometimes called axial) and the radial as well. Later, rather than particle velocities, pressure distribution was evaluated by others; pressure distribution, of course, can be easily translated into a particle velocity field.

8.1 Hard-Wall Bends

Velocity or pressure distribution across bends is calculated by integrating the basic equations of motion and using appropriate boundary conditions to evaluate the constants of integration. Except for Rostafinski, all work in this area was done by numerical integration and thus there is limited insight into the parametric dependence of those distributions on boundary conditions. Since Rostafinski (1972) limited his analytical scope to extremely low frequencies, this particular contribution pertains only to propagation of long waves. Nevertheless, because his results are based on analytical derivations, some elements of his analysis are now given.

Tangential particle velocity in a bend at $\theta = 0$, that is, at the bend inlet, where the vibrational displacements $v_0\, e^{(i\omega t)}$ of a piston generate harmonic waves, equals

$$\frac{1}{r}\frac{\partial \phi}{\partial \theta}\bigg|_{\theta=0} = \sum_{m=0}^{\infty} \frac{1}{r} \frac{C_m \nu_m}{\sin(\pi \nu_m)} F_{\nu_m} = v_0$$

where

$$F_{\nu_m} = J'_{\nu_m}(kR_1) J_{-\nu_m}(kr) - J_{\nu_m}(kr) J'_{-\nu_m}(kR_1)$$

as given before (eq. (6.2)). Using the orthogonality conditions satisfied by the set F_{ν_m} gives

$$\frac{C_m}{\sin(\pi \nu_m)} = \frac{\displaystyle\int_{R_1}^{R_2} v_0 F_{\nu_m} dr}{\nu_m \displaystyle\int_{R_1}^{R_2} \frac{1}{r} F_{\nu_m}^2 dr}$$

The integral in the denominator is easily reducible to a sum of integrable expressions of which one vanishes. The integral

$$\int_{R_1}^{R_2} v_0 F_{\nu_m} dr$$

can be integrated for any desired vibration at $\theta = 0$. This has been done for three boundary conditions, as shown in table 8.1.

The first boundary condition, that the particle velocity at $\theta = 0$ at the piston is independent of radius, yields amplitudes for the basic mode that depend on the radius ratio $a = R_2/R_1$ and evanescent waves that depend partly on this ratio as well. The second boundary condition (v_0 is proportional to radius) results in the amplitude of the wave and of the decaying vibrations strongly increasing with a. The third boundary condition (potential vortex at the inlet) results in a wave of almost constant amplitude, independent of a

TABLE 8.1.—EFFECT OF RADIAL DISTRIBUTION OF INITIAL VELOCITY v_0

[From Rostafinski (1970).]

Initial particle velocity distribution	Amplitude of progressing wave proportional to –	Amplitude of mth decaying wave proportional to –
$v_0 = $ const.	$\dfrac{a-1}{\ln a} + O(kr)^2$	$(a \cos m\pi - 1) + O(kr)^2$
$v_0 \dfrac{r}{R_1}$	$\dfrac{a^2 - 1}{2 \ln a} + O(kr)^2$	$2\left(\dfrac{m^2\pi^2 + (\ln a)^2}{m^2\pi^2 + 4(\ln a)^2}\right)(a^2 \cos m\pi - 1) + O(kr)^2$
$v_0 \dfrac{R_1}{r}$	$1 + O(kr)^2$	$0 + O(kr)^2$

and with vanishingly small damped oscillations at the inlet. The assumption that the tangential velocity at $\theta = 0$ is inversely proportional to the radius results in a remarkable simplification of the equation of motion because virtually only the undamped simple propagating and weak radial waves are present. Furthermore, the propagating waves are essentially independent of the bend radius ratio.

The first boundary condition ($v_0 = $ constant) and the third have been simulated on a water table (fig. 8.1). Tests were run by Rostafinski at the von Karman Institute for Fluid Dynamics in Belgium, in 1965. Two curved sheet metal bands simulated a cylindrical bend; waves were generated by rolling a 30-mm-diameter plastic rod back and forth on the table at the bend inlet. Analysis of the first boundary condition ($v_0 = $ constant) showed that a system of reflections (radial waves) characterizes the propagation of plane waves ($v_0 = $ constant) induced at the bend inlet. For the third boundary condition, the potential vortex $[v_0 = f(1/r)]$ at the inlet, the motion is greatly simplified; virtually all evanescent waves at the inlet are eliminated.

Analysis indicates that initially plane waves propagating into curved ducts are profoundly influenced by the curvature of the walls. The tangential particle velocities are (for extremely low frequencies) almost exactly inversely proportional to the radius, that is, velocities follow the distribution of the potential vortex. It is thus obvious that pressure is almost constant at any wave front. Figure 8.2 shows six curves calculated for kR_1 of 0.02 and 0.03.

The radial particle velocities are approximately two orders of magnitude smaller than the tangential particle velocities. Their radial distribution is governed by squares of kR_1 and kr, which are small. Figure 8.3 shows typical curves of the standing radial particle velocities. Their shape is characterized by lack of symmetry: the maxima of the curves are shifted toward the bend inner wall.

The attenuated tangential, evanescent particle velocities characterize the transition of plane waves at the bend inlet to shapes proper for cylindrical geometries. Figure 8.4 gives results of a sample calculation illustrating the behavior of those oscillations for a duct radius ratio of 2. The vibrations are basically of low amplitude. Even at the position $\theta = \pi/16$, close to the vibrating piston face at $\theta = 0$, they are one order of magnitude smaller than the nonattenuated propagating velocities. The radial distribution of these oscillations changes significantly with wave angular position in the duct. At $\theta = \pi/4$ these oscillations are reduced to a low level and are nearly uniform across the duct width. Figure 8.5 shows the same oscillations calculated for three duct radius ratios but for a single angular position of $\theta = \pi/4$. It indicates that the decaying, evanescent oscillations are much more pronounced and extend farther when induced in a wider duct.

The nonpropagating, evanescent radial particle velocities at the duct inlet are illustrated in figure 8.6. Note that the asymmetry of these vibrations is much more severe than that of the propagating radial particle velocities (fig. 8.2). Furthermore, the evanescent components are about twice as strong as the propagating components of the particle velocity fields.

After the publication of Rostafinski's (1972) data on long waves, Cummings (1974) published a much needed development on shorter wavelengths and above all provided the first experimental verification of the calculated characteristics. Except for recalculated coordinates, graphs given by Cummings (1974) are reproduced here as published (figs. 8.7 and 8.8). They illustrate the excellent agreement of his numerical calculations with measurements.

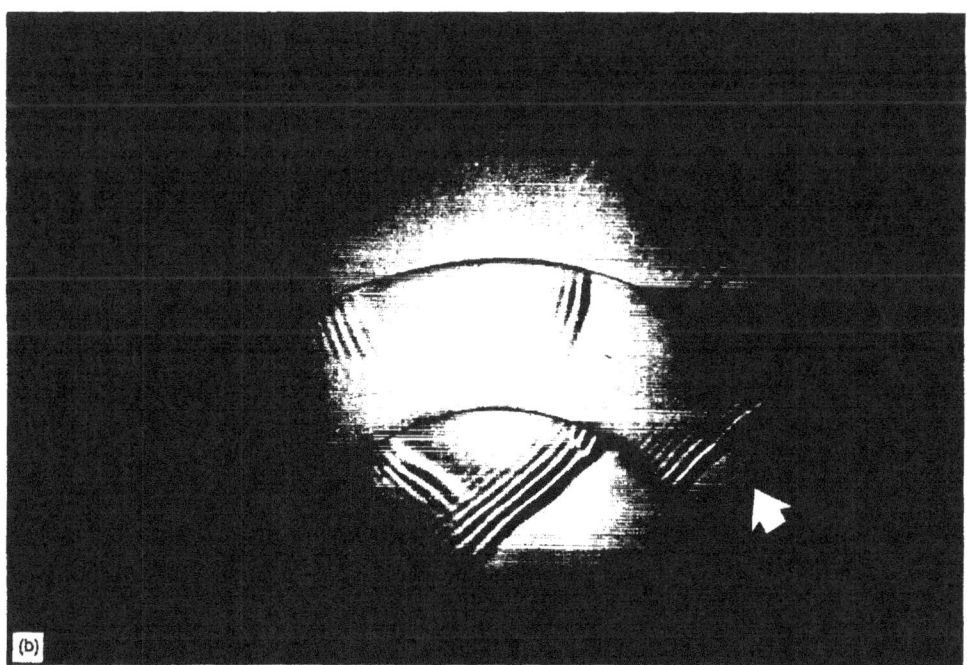

(a) Constant particle velocity at inlet, v_0.

(b) $v_0 = f(1/r)$.

Figure 8.1.—Simulation on a water table of first and third cases in table 8.1.

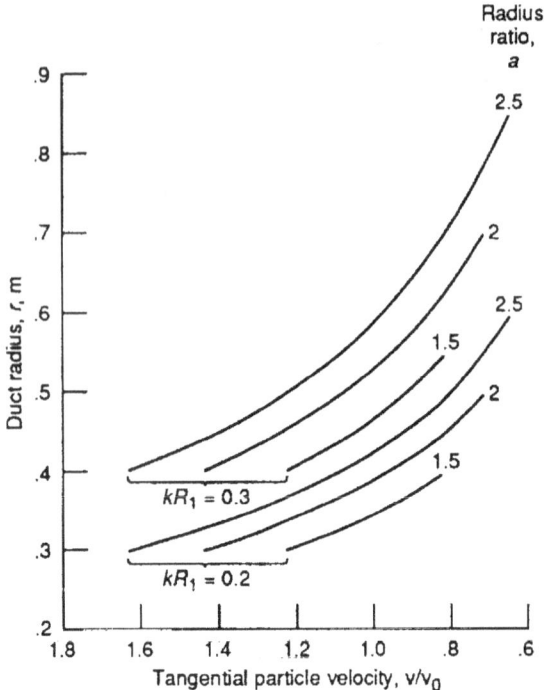

Figure 8.2.—Tangential particle velocities of propagating long waves in curved
, ducts. From Rostafinski (1972).

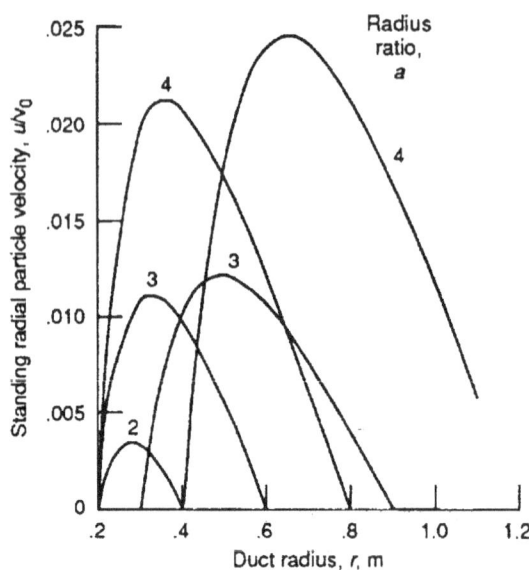

Figure 8.3.—Standing radial particle velocities in bends for three radii of
convex (inner) wall of bend, three wave number parameters kR_1, and three
radius ratios. From Rostafinski (1972).

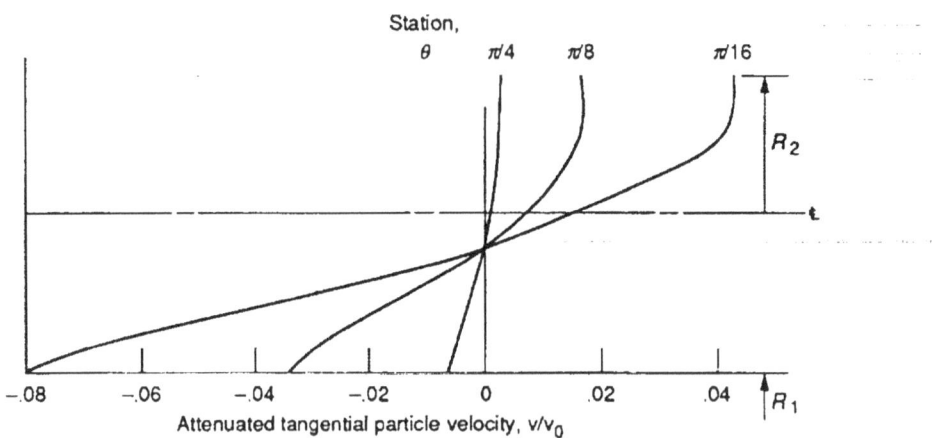

Figure 8.4.—Attenuated tangential particle velocities for three angular positions in bend. Radius ratio, $a = 2$. From Rostafinski (1972).

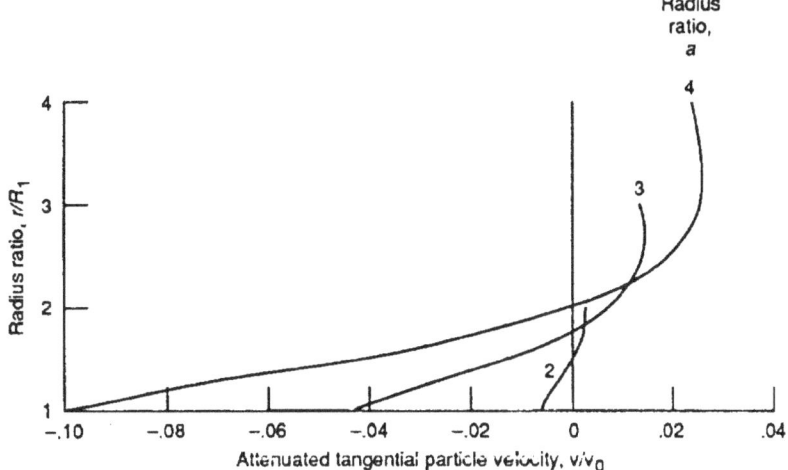

Figure 8.5.—Attenuated tangential particle velocities for three radius ratios at angular position $\theta = \pi/4$. From Rostafinski (1972).

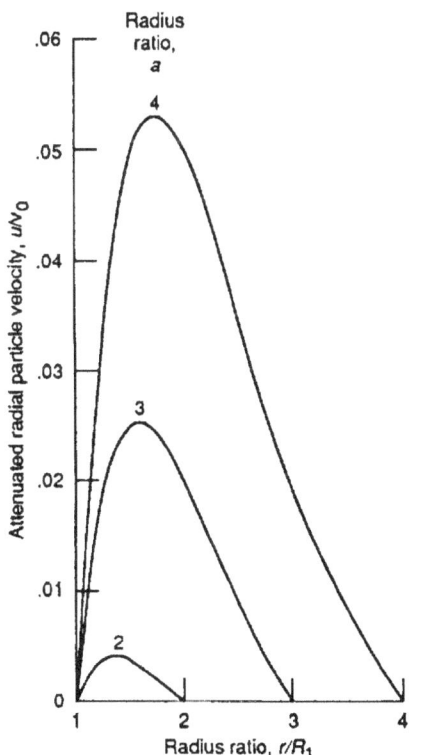

Figure 8.6.—Attenuated radial particle velocities at angular position $\theta = \pi/4$. From Rostafinski (1972).

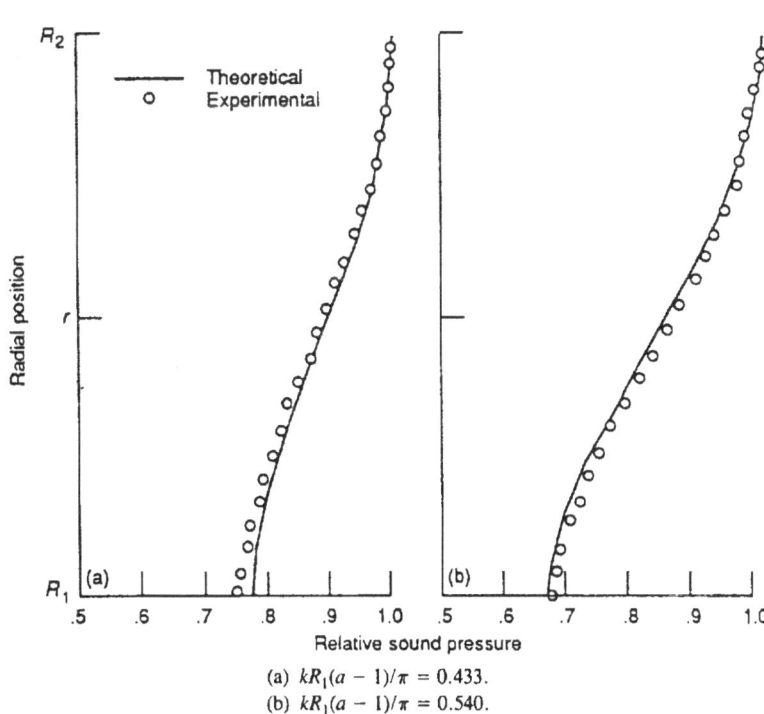

(a) $kR_1(a - 1)/\pi = 0.433$.
(b) $kR_1(a - 1)/\pi = 0.540$.

Figure 8.7.—Sound field in bend I for two wave number parameters $kR_1(a - 1)/\pi$. Radius ratio, $a = 1.587$. From Cummings (1974).

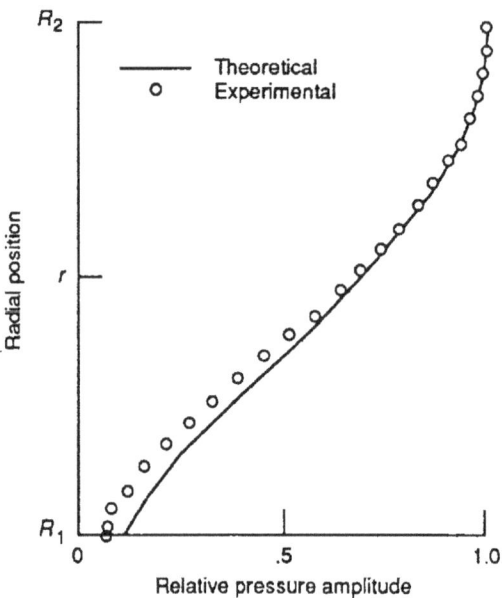

Figure 8.8.—Sound field in bend II. Wave number parameter, $kR_1(a - 1)/\pi = 0.67$; radius ratio, $a = 10.3$. From Cummings (1974).

Figure 8.7 shows the radial distribution of sound pressure for the propagating mode (0,0) in a bend with a radius ratio a of 1.587 and wave number parameters kR_1 of 2.316 and 2.888. Figure 8.8 shows data on the propagating mode (0,0) in a sharp bend with $a = 10.3$ and $kR_1 = 0.228$. As noted by Cummings the variation in the acoustic pressure amplitude across the duct becomes more marked with increasing frequency. From figure 8.8 we learn that with increasing bend sharpness the pressure distribution nonuniformity of the propagating mode becomes much more pronounced.

Figure 8.9 shows pressure and velocity fields for a bend with $a = 2$. The data are taken from Rostafinski's (1972) paper (on extremely low frequency waves) and recalculated by Cummings to emphasize the pressure distribution of both propagating and evanescent waves in curved ducts. Finally in figure 8.10 Cummings compares the pressure and tangential particle velocities of the propagating mode (0,0) in a bend with $a = 10.3$ and $kR_1 = 0.228$. Here the radial pressure gradient is much more marked than in figure 8.9.

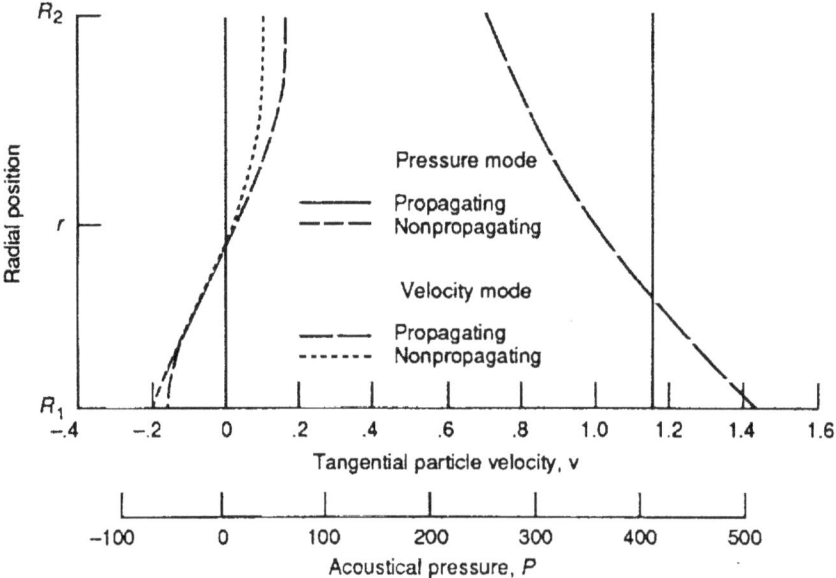

Figure 8.9.—Pressure and velocity fields for bend with radius ratio $a = 2$. Wave number parameter, $kR_1(a - 1)/\pi = 0.0064$ (calculated from Rostafinski, 1972). From Cummings (1974).

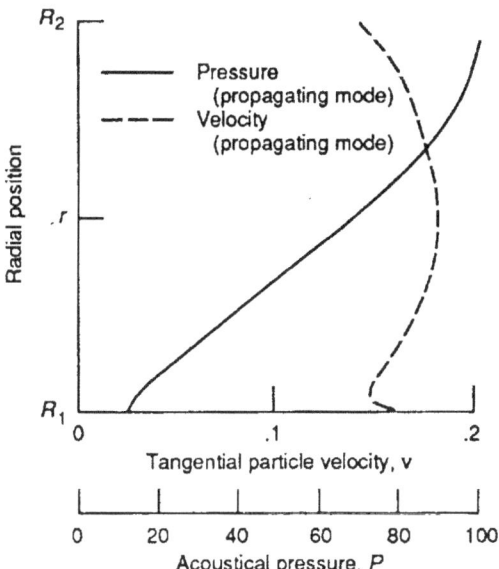

Figure 8.10.—Pressure and velocity fields for bend II. Wave number parameter, $kR_1(a - 1)/\pi = 0.67$; radius ratio, $a = 10.3$. From Cummings (1974).

In two analytical studies Rostafinski (1974a, 1976) analyzed pressure and particle velocities in bends with several radius ratios and in a range of higher frequencies. To bypass the difficulty of solving the characteristic equation for Bessel functions of large-valued fractional order, he used an inverse method. Since the Bessel functions of order $(m + 1/2)$, $m = 0,1,2,...$ can be written in a simple, closed form, the characteristic equation was solved not for (the imposed) angular wave numbers $(m + 1/2)$ but for frequencies as given by the Bessel function argument kR_1. Table 8.2 gives the wave number parameters kR_1 that satisfy the boundary conditions (hard walls) for the imposed angular wave numbers $(m + 1/2)$. The objective of this work was not finding an algorithm but studying the characteristics of propagation in a higher frequency range. Figures 8.11 and 8.12 give several of Rostafinski's results pertaining to the propagating modes. As indicated in figure 8.11, higher frequencies induce significant changes in tangential particle velocity distributions across ducts. Figure 8.12(b) shows both tangential and radial particle velocity distributions for higher modes.

The redistribution of particle velocities and acoustic pressure with the angular wave position is shown in figures 8.13 and 8.14. Radial distribution of the tangential particle velocities strongly depends on frequency.

TABLE 8.2.—PROPAGATION
PARAMETERS FOR RANGE
OF ANGULAR WAVE
NUMBER

[Radius ratio, $a = 2$;
from Rostafinski
(1976).]

Wave number parameter, kR_1	Angular wave number, ν_m
0.3396	0.5
1.0115	1.5
1.6633	2.5
2.2869	3.5
2.8817	4.5
3.4523	5.5
4.0065	6.5

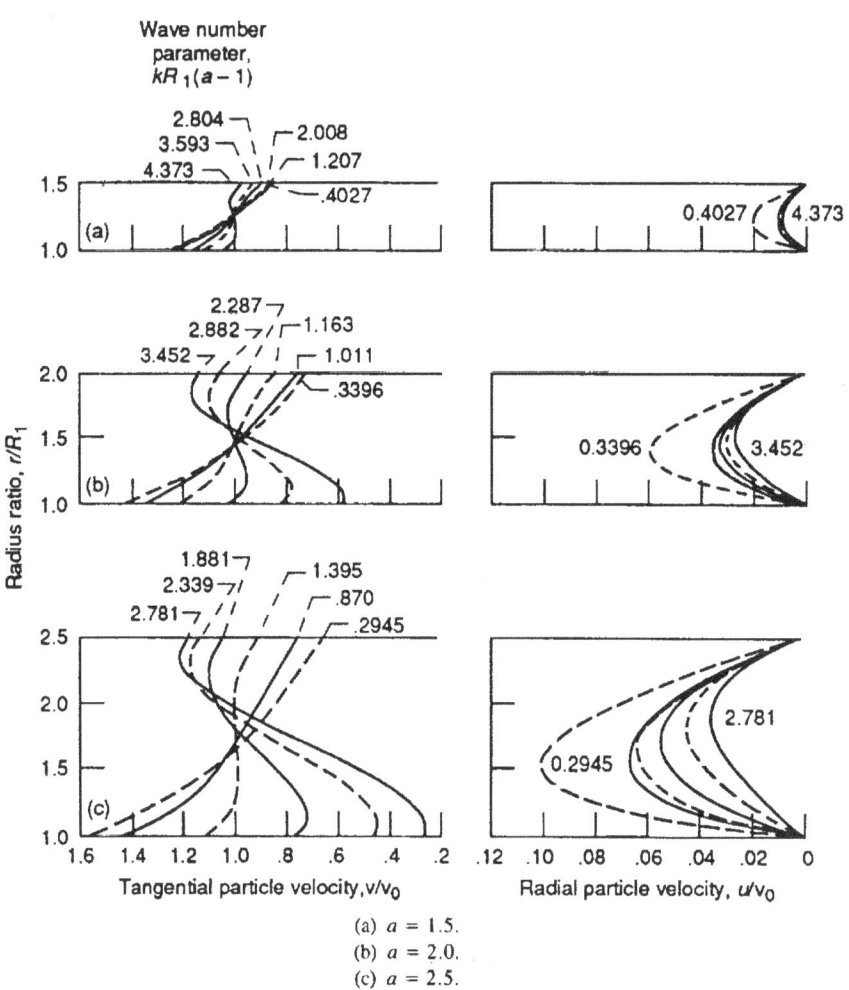

Figure 8.11.—Particle velocities in three ducts of different radius ratios a with radius of convex (inner) wall of bend $R_1 = 0.2$ m. (The solid and dashed lines are used for easier identification of curves only.) From Rostafinski (1974a).

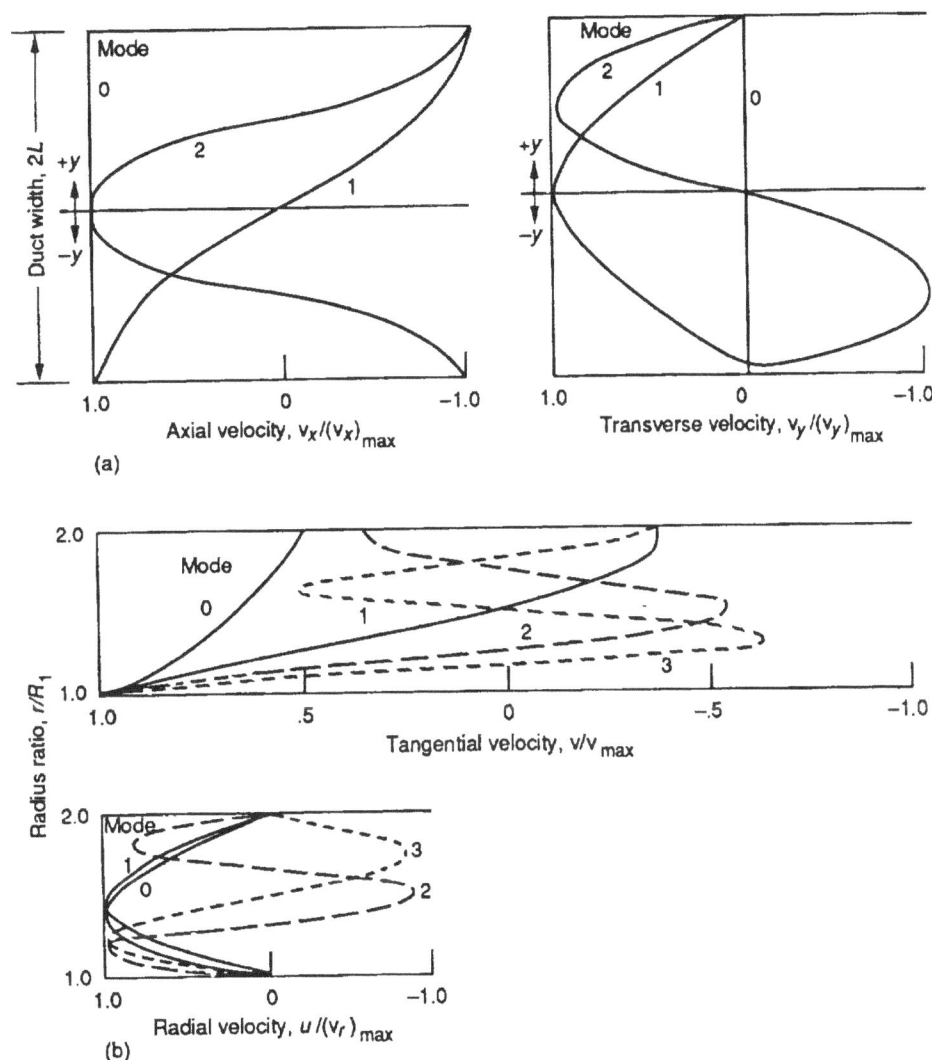

(a) In straight duct for one frequency and three consecutive modes.
(b) In curved duct for $\nu_i = 0.5$ for four consecutive modes.

Figure 8.12.—Particle velocities in a curved and a straight duct. All curves are normalized to maximum amplitude of 1.0. From Rostafinski (1974a).

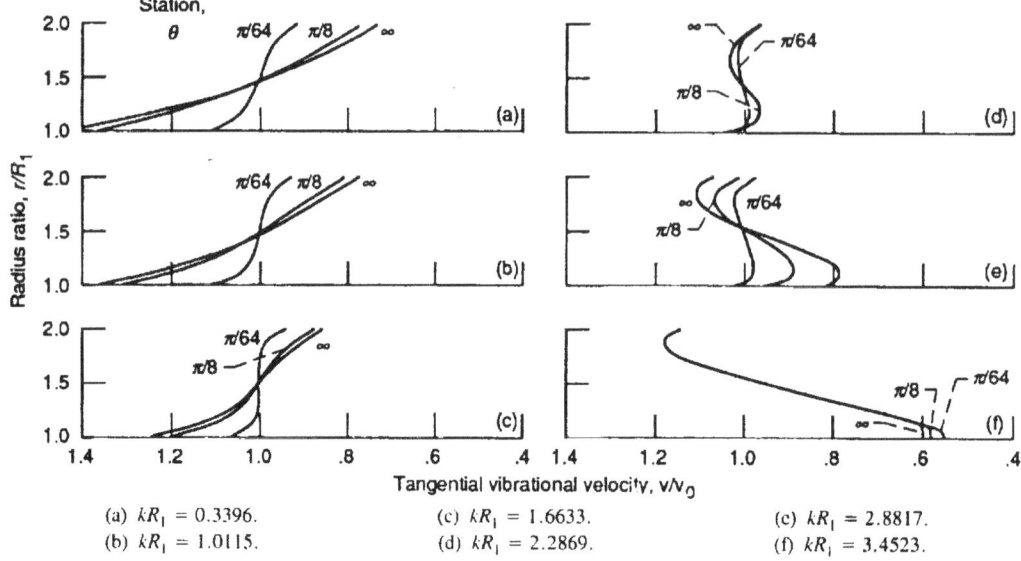

Figure 8.13.—Tangential particle velocities in bends for six wave number parameters kR_1. From Rostafinski (1976).

(a) $kR_1 = 0.3396$.
(b) $kR_1 = 1.0115$.
(c) $kR_1 = 1.6633$.
(d) $kR_1 = 2.2869$.
(e) $kR_1 = 2.8817$.
(f) $kR_1 = 3.4523$.

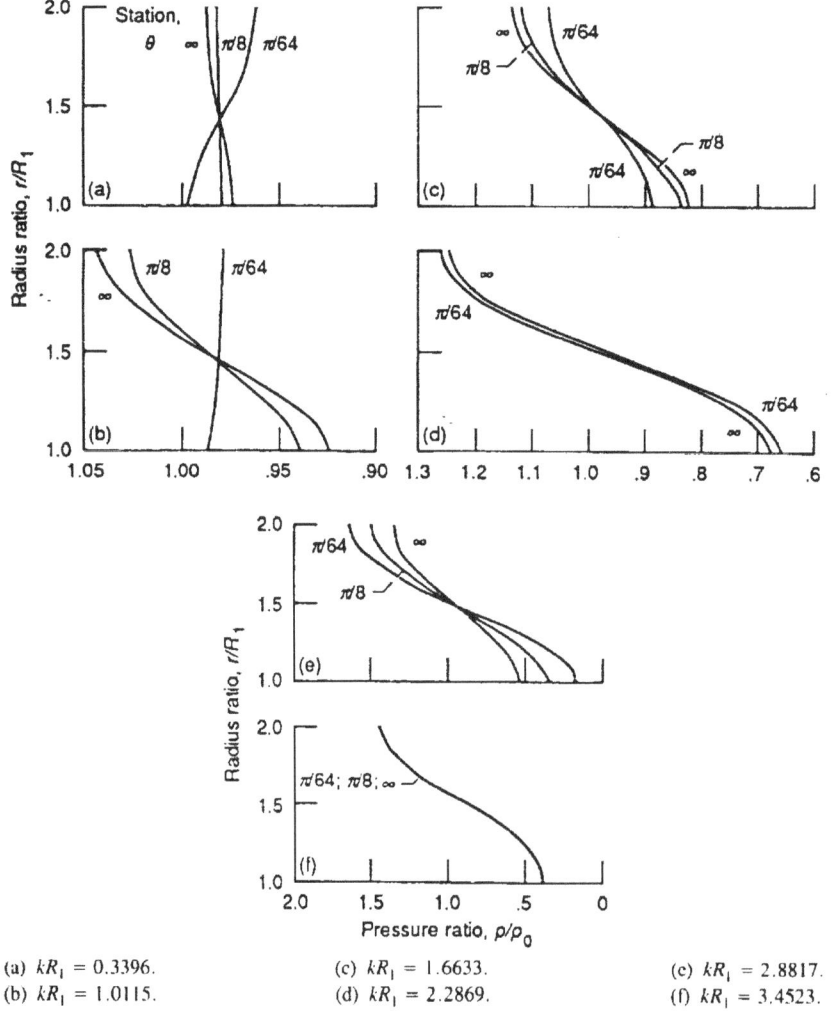

(a) $kR_1 = 0.3396$.
(b) $kR_1 = 1.0115$.
(c) $kR_1 = 1.6633$.
(d) $kR_1 = 2.2869$.
(e) $kR_1 = 2.8817$.
(f) $kR_1 = 3.4523$.

Figure 8.14.—Acoustic pressure in bends for six wave number parameters kR_1. From Rostafinski (1976).

Attenuation of the nonpropagating modes plays an important role in this distribution. At higher frequencies the "steady state" distribution sets in early, close to the piston face. Acoustic pressure distribution is barely noticeable at extremely low frequencies. It stabilizes quite early at higher frequencies.

Myers and Mungur (1976) used numerical techniques to obtain pressure distribution curves for three propagating modes. Their graph, with coordinates modified to conform to the units adopted in this monograph, is shown in figure 8.15. It supplements data in figure 8.12.

Fuller and Bies (1978b) give two graphs (fig. 8.16) showing the theoretical variation in pressure amplitude across the duct. It was evaluated by using the following expression:

$$\text{Relative sound pressure} = \frac{(F_{00})_r}{(F_{00})_{R_1}}$$

where F_{00} is the characteristic function of the (0,0) mode. Two frequencies have been used, corresponding to $kR_1(a - 1)$ of 1.5 and 3.0 calculated for two ducts ($a = 1.28$ and $a = 2.25$). The experimental data in figure 8.16(b) indicate that the theoretical curve is indeed verified. It can be seen again, as stated by the authors, that at low frequencies, corresponding to $kR_1(a - 1) < 1.5$, acoustic waves propagate with only a small variation in pressure amplitude across the duct for the bends shown in figure 8.16. Hence, at low frequencies the basic design assumption that acoustic waves propagate as plane waves in all sections of a bend leads to negligible error. However, as the frequency increases, the variation in sound pressure amplitude across the duct becomes more pronounced, particularly in the sharper bend. Thus, for sharp bends at high frequencies the assumption of plane waves does not hold. This has been confirmed experimentally by Cummings (1974).

Fuller and Bies (1978b) further recalled that Rostafinski (1972) had shown that the tangential velocity distribution for the (0,0) mode follows closely that of a potential vortex for low wave number parameters ($kR_1 < 1$) and is close to that of a forced vortex at higher wave number parameters ($kR_1 = 3$).

The distribution of the particle velocities and pressures for the nonpropagating, evanescent, higher modes has been analyzed by Rostafinski (1976), also by using the "inverse method" of solving the characteristic equation. The calculated nonpropagating tangential and radial particle velocities at two tangential positions, at $\theta = \pi/64$ (close to the face of the vibrating piston) and at $\theta = \pi/4$, are shown in figures 8.17 and 8.18. Besides the distribution of velocities, these figures also show the degree of attenuation, which strongly depends on wave number parameter kR_1. Note that at some kR_1 the evanescent vibrations may be negative.

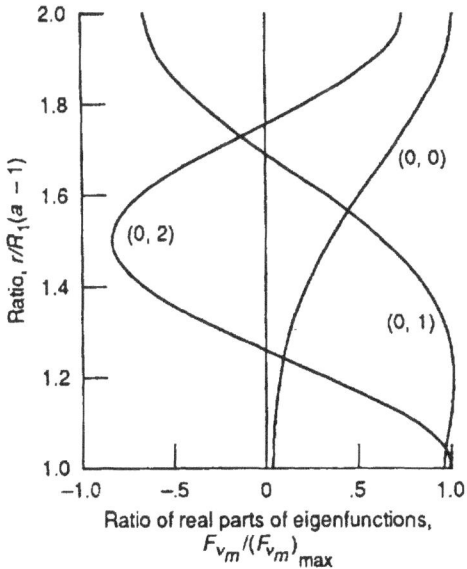

Figure 8.15.—Pressure mode shapes for rigid-wall duct. Nondimensional wall admittance, $\eta = 0$; radius ratio, $a = 2$; wave number parameter, $kR_1(a - 1) = 6$. From Myers and Mungur (1976).

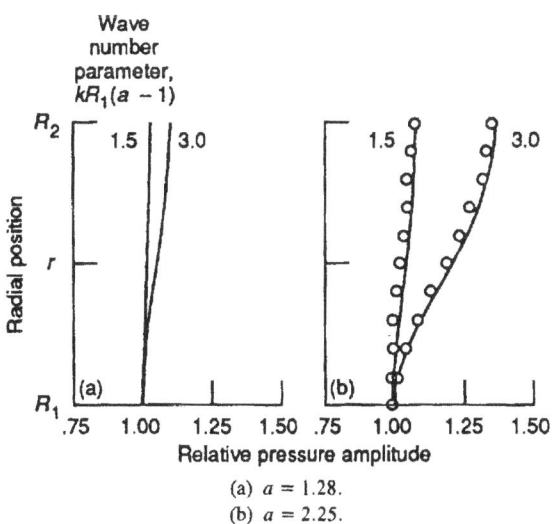

(a) $a = 1.28$.
(b) $a = 2.25$.

Figure 8.16.—Variation in pressure amplitude for bends with two radius ratios a. From Fuller and Bies (1978b).

27

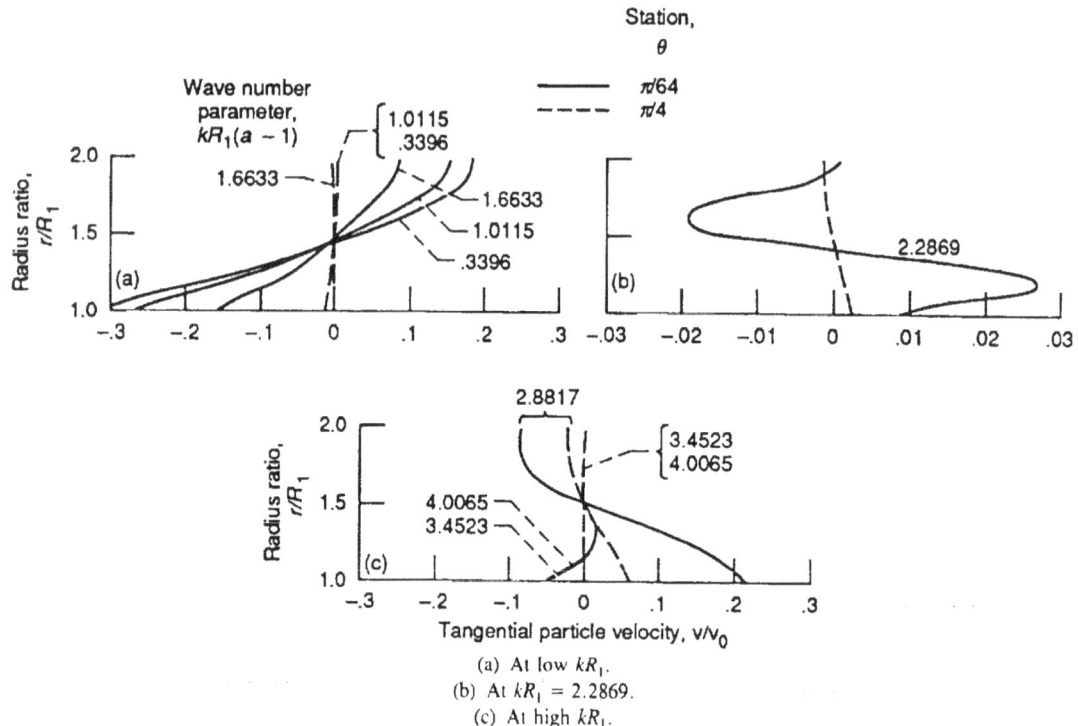

(a) At low kR_1.
(b) At $kR_1 = 2.2869$.
(c) At high kR_1.

Figure 8.17.—Nonpropagating tangential particle velocities at various wave number parameters kR_1. From Rostafinski (1976).

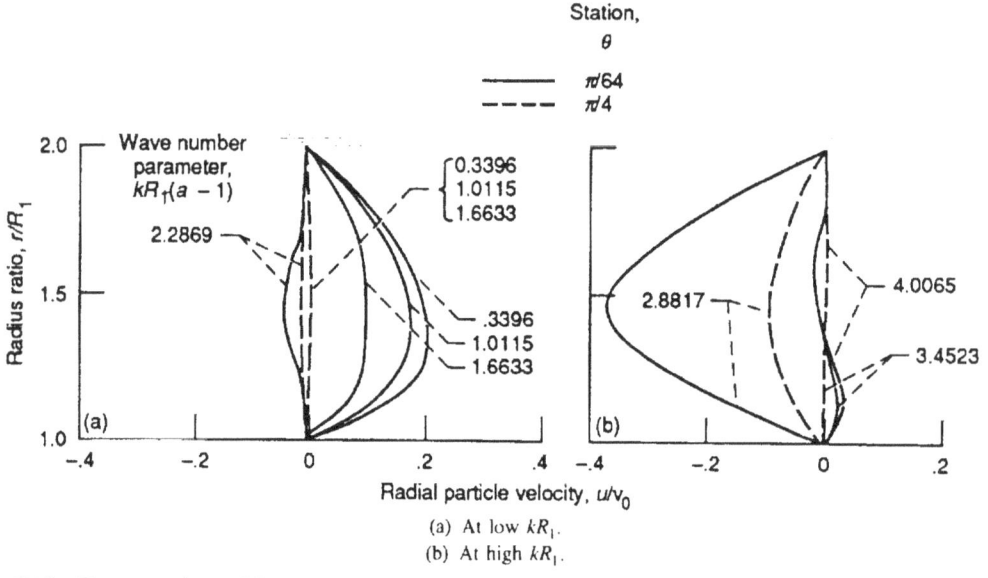

(a) At low kR_1.
(b) At high kR_1.

Figure 8.18.—Nonpropagating radial particle velocities at high and low wave number parameters kR_1. From Rostafinski (1976).

A paper by El-Raheb and Wagner (1980) exhibits extensive normalized isobars at resonant frequencies for a 90° bend with $a = 5$. These data are given in section 10.0.

Interesting data have been published by Rostafinski (1972), Osborne (1976), and Cabelli (1980) showing particle velocities and acoustic pressures in bends connected to straight ducts and provided with anechoic terminations so that the end reflections are eliminated. Rostafinski gives particle velocities at the bend's inner and outer walls for two situations: an infinite bend (a coil) and a bend followed by a straight duct. This graph (fig. 8.19) shows particle velocity differences at the two curved walls and their rate of change at the bend inlet and outlet and in the straight duct that follows the bend, where the wave becomes plane again. These theoretical evaluations for extremely low frequencies and for a bend–straight duct system have been simulated by Rostafinski on a water table (fig. 8.20). The plane wave at the bend inlet is reflected from the outer wall and proceeds to the outlet. In the straight duct it is re-formed almost to its original plane wave configuration. Extensive studies, both analytical and experimental, of the pressure distribution were done by Osborne (1976) and Cabelli (1980). Osborne, besides various characteristics of acoustic transmission in bends, calculated the pressure field by the separation of variables, writing equations for continuity of pressure and particle velocity at discontinuities (junctions) and solving them by numerical integration. He checked the analytical results by testing the sound field in a bend joining two straight ducts.

Figure 8.21 gives the geometry and location of the measurement points of Osborne's experimental setup, a 45° bend with $a = 2$. Figures 8.22 to 8.24 give his analytical and experimental results. In general, his theoretical predictions have been closely verified by tests. Interestingly, his data pertain to high-frequency regions: 630 to 1000 Hz. All sound sources except one are of the (0,1) mode. At 800 Hz both modes (0,1) and (0,0), the plane wave, have been evaluated. His angular wave numbers ν range up to 15.5 for the (0,0) mode and up to 11.0 for the (0,1) mode; this represents propagation of the (0,0), (0,1), (0,2), (1,0), and (1,1) modes in the bend.

Osborne (1976) gives the following description of his figures:

> In each figure the predicted distribution at the start of the bend due to the sound pressure field in the first straight duct is indicated by the chain dotted line whilst that calculated from the modal constants…is indicated by the full line. At the end of the bend the sound field calculated from the bend modal constants is represented by the broken line and that from the final straight duct modal constants…by the full line. The differences between the two sets of calculated values at each bend junction are slight and may be the result of ignoring the contributions of the evanescent modes….

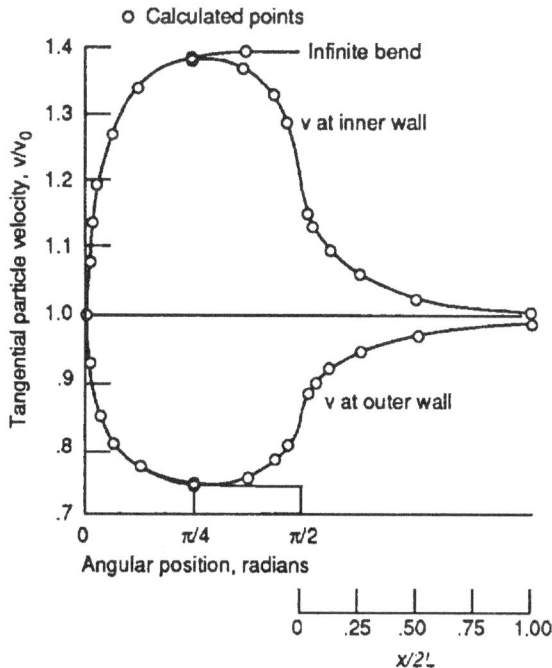

Figure 8.19.—Propagation in bend–straight duct system. Tangential particle velocities on walls. From Rostafinski (1972).

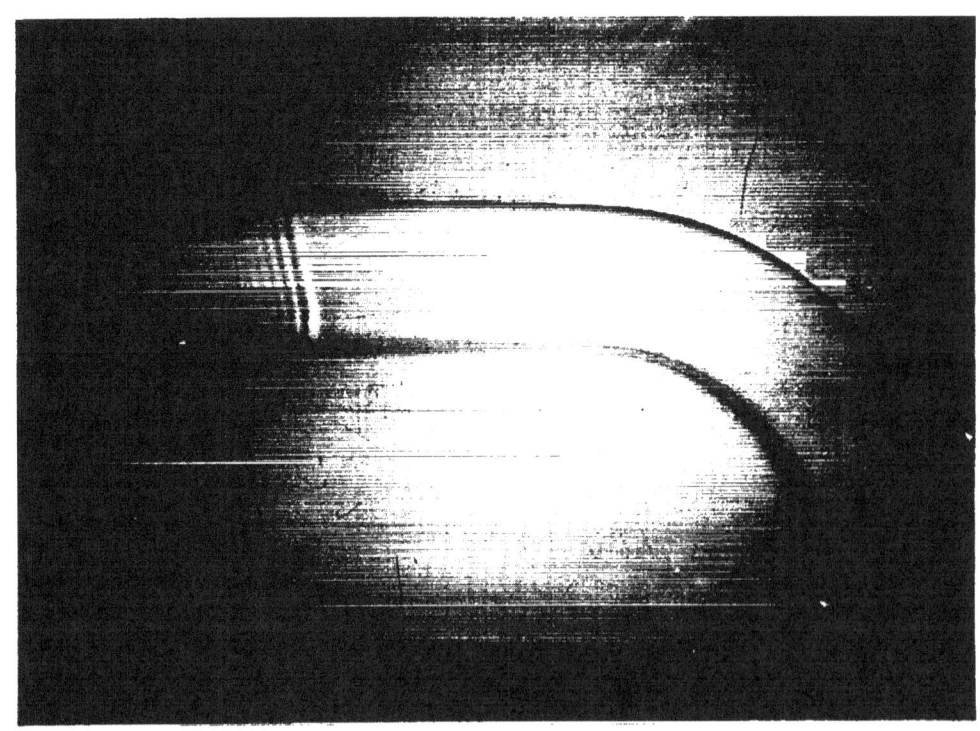

Figure 8.20.—Simulation on a water table of plane wave propagation in bend followed by straight ducts.

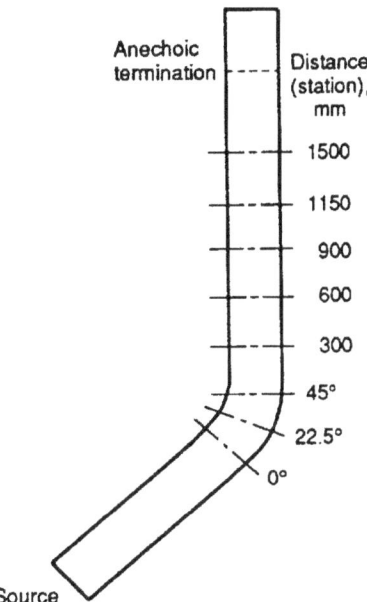

Distance
(station),
mm

1500

1150

900

600

300

45°

22.5°

0°

Source

Figure 8.21.—Plan of test ducts showing measurement sections. From Osborne (1976).

Figure [8.24] shows contours of sound pressure level interpolated from traverses on sections midway between the non-curved walls of the bend.... The total number of spot readings being of the order of 350 at each frequency.

Cabelli (1980) evaluated a 45° and a 180° bend of radius ratio $a = 9$ at two frequencies corresponding to kR_1 of 0.31 and 0.375. His measurements give pressure amplitude in various locations in a bend and in the two straight ducts connected to the bend inlet and outlet. In his analysis the equations and the boundary conditions were approximated by finite difference equations applied at the nodes of a mesh superimposed on the solution domain. Second-order central differences were used to approximate derivatives. The obtained set of simultaneous equations was expressed in partitioned matrix form as described by Baumeister and Rice (1975).

The results of Cabelli's studies are summarized in several figures. Figure 8.25 shows deviations from initially uniform pressure distribution in upstream and downstream straight ducts; consequently, Cabelli proceeds with modal analysis of the sound field. The results are given in figure 8.26. His comments on this figure are as follows:

...Only the plane wave and the first two higher order modes were significant and are featured in the figure. The small variation in the amplitude of the plane wave downstream of the bend is typical of the errors which were described in relation to the propagation and the decay in a straight duct. The upstream variation of this component, however, is significant. It represents the standing wave pattern associated with the reflection of sound from the discontinuity. The reflection coefficient for $k = 2.5$ was equal to 0.15 approximately. In addition to the reflection of the propagating plane wave, the bend was responsible for the presence of the higher order modes. As expected, however, these decayed exponentially. The exponent of the rate of decay for the first order mode was found from the numerical results to be approximately equal to 1.8. Analysis predicts a value of 1.9. For higher order modes the discrepancy was greater; numerical results indicated decay rates of 4.9 and 6.8 for the second and third orders whereas analysis predicts corresponding rates of 5.8 and 9.1. The magnitude of these modes was sufficiently small, however, that the errors would have a negligible influence on the pressure distribution. Figure [8.27] also shows the standing wave pattern for the plane wave when $k = 3.0$. The reflection coefficient in this case was found to be approximately equal to 0.4. The corresponding distribution of the pressure amplitude across the duct is shown at selected sections through the system. The greater distortion of the pressure profile is evidence of the increasing magnitude of the evanescent higher orders. This aspect is reviewed later.

The characteristics of the 45° bend were studied numerically over a range of values of k less than π.

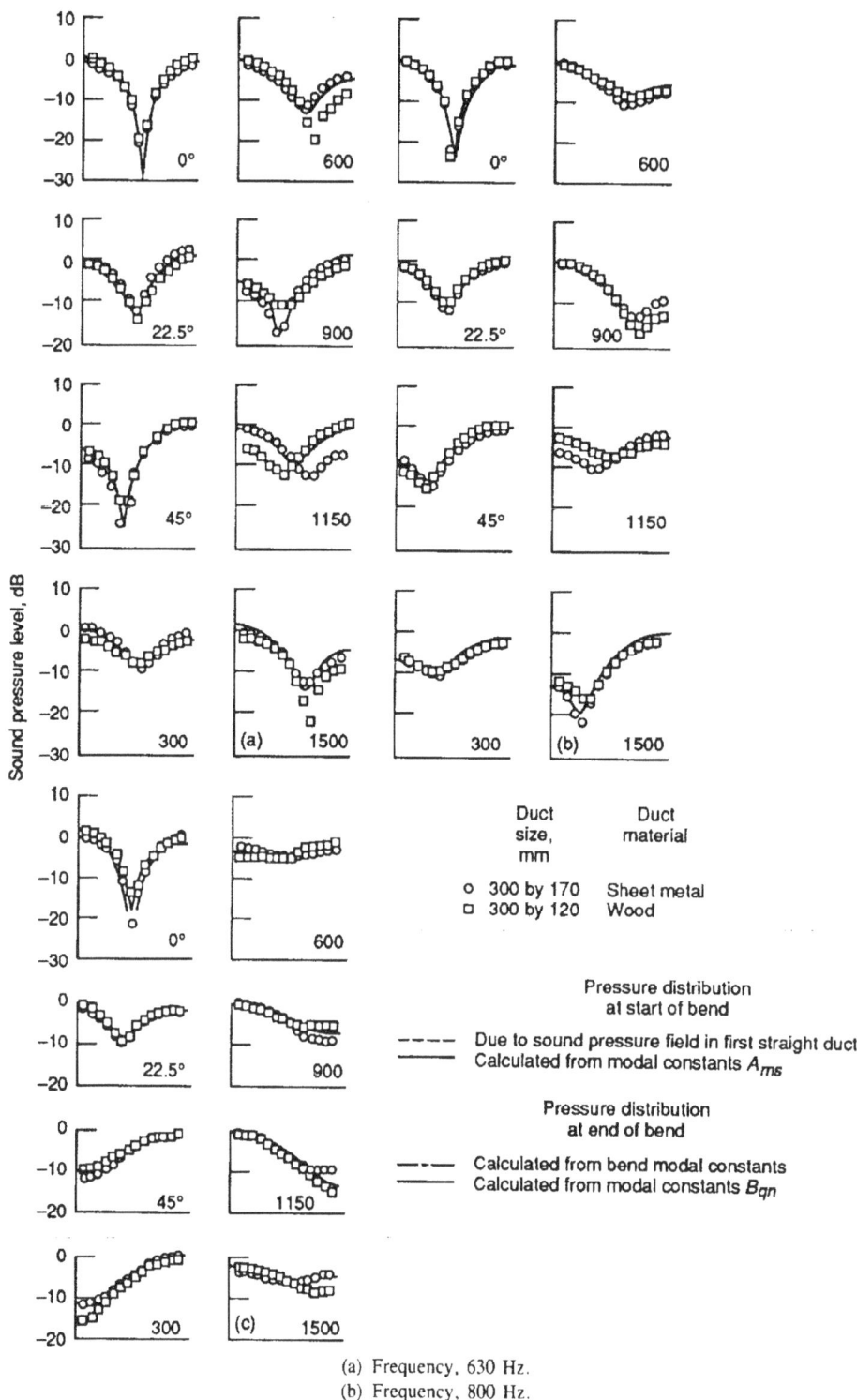

(a) Frequency, 630 Hz.
(b) Frequency, 800 Hz.
(c) Frequency, 1000 Hz.

Figure 8.22—Measured and predicted sound pressure level distributions in duct system due to first cross-mode source in upstream straight section. From Osborne (1976).

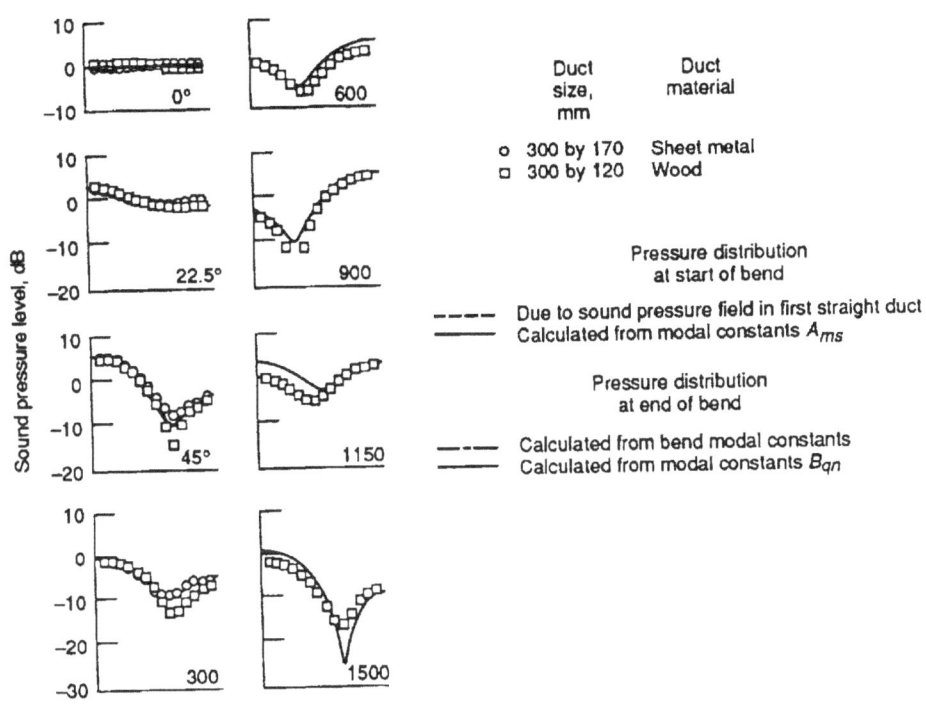

Figure 8.23.—Measured and predicted sound pressure level distributions in duct system due to plane wave source in upstream straight section at 800 Hz. From Osborne (1976).

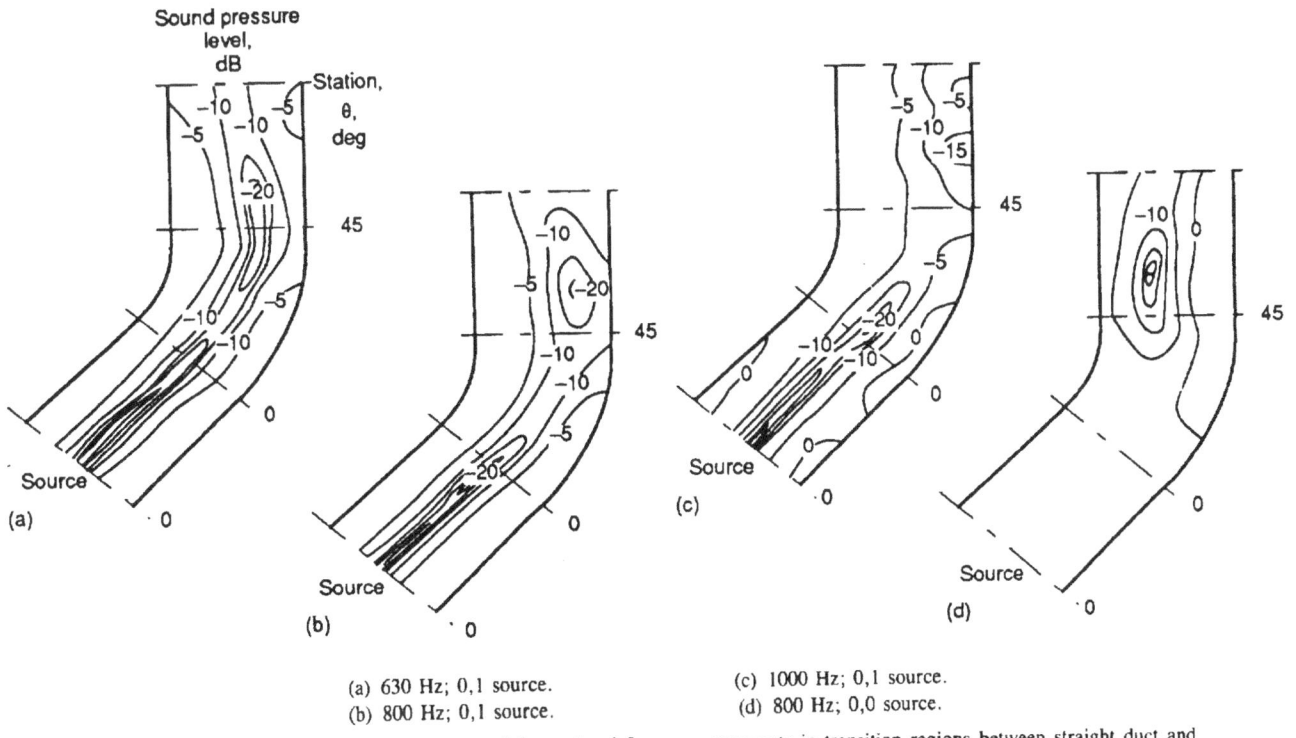

(a) 630 Hz; 0,1 source.
(b) 800 Hz; 0,1 source.

(c) 1000 Hz; 0,1 source.
(d) 800 Hz; 0,0 source.

Figure 8.24.—Contours of sound pressure level interpolated from measurements in transition regions between straight duct and bend for various frequencies and sources. From Osborne (1976).

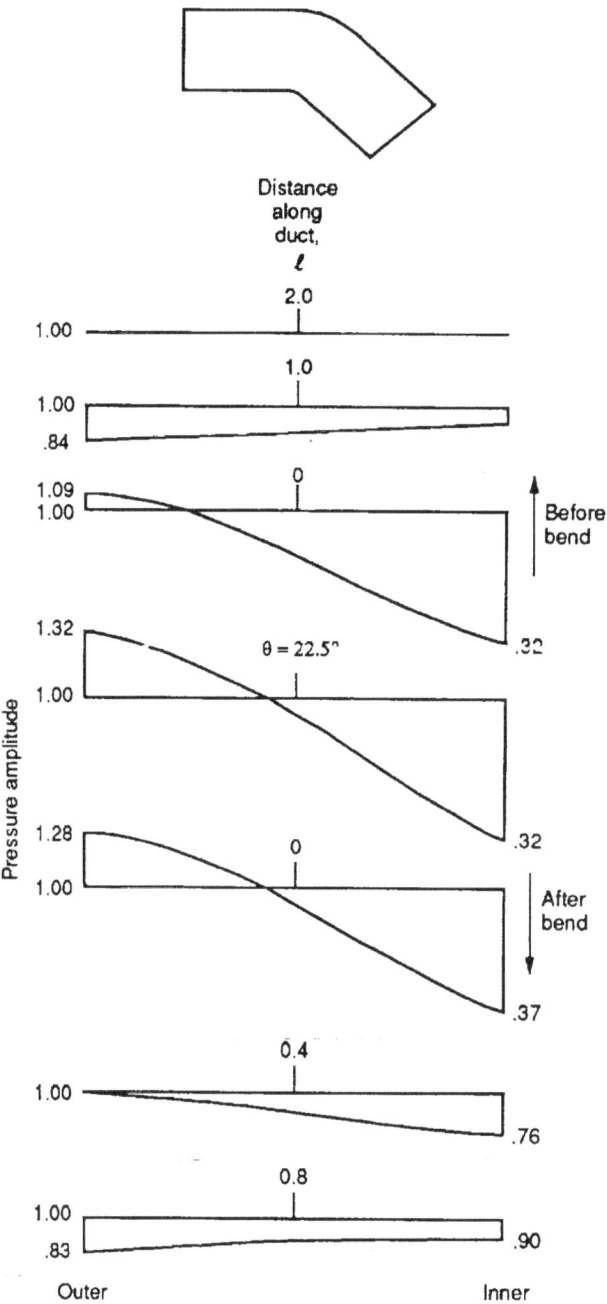

Figure 8.25.—Distribution of pressure amplitude across duct at selected sections for a 45° bend. Wave number parameter, $kR_1(a - 1) = 2.5$; radius ratio, $a = 9$. From Cabelli (1980).

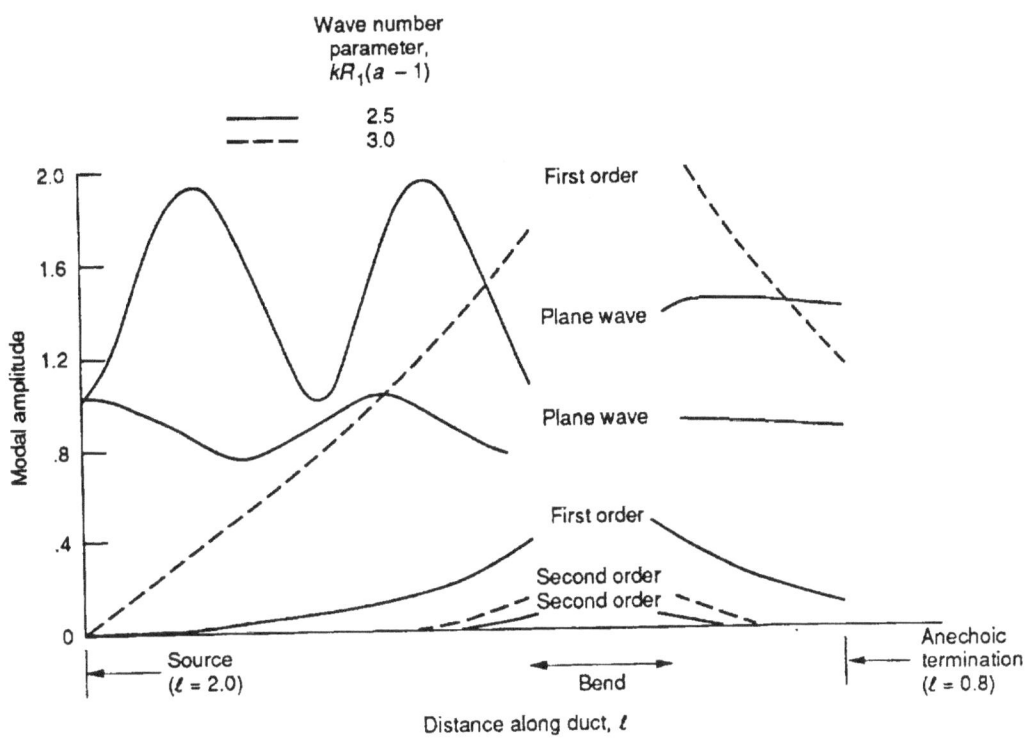

Figure 8.26.—Distribution of modal amplitudes along straight sections of duct system and a 45° bend. Radius ratio, $a = 9$. From Cabelli (1980).

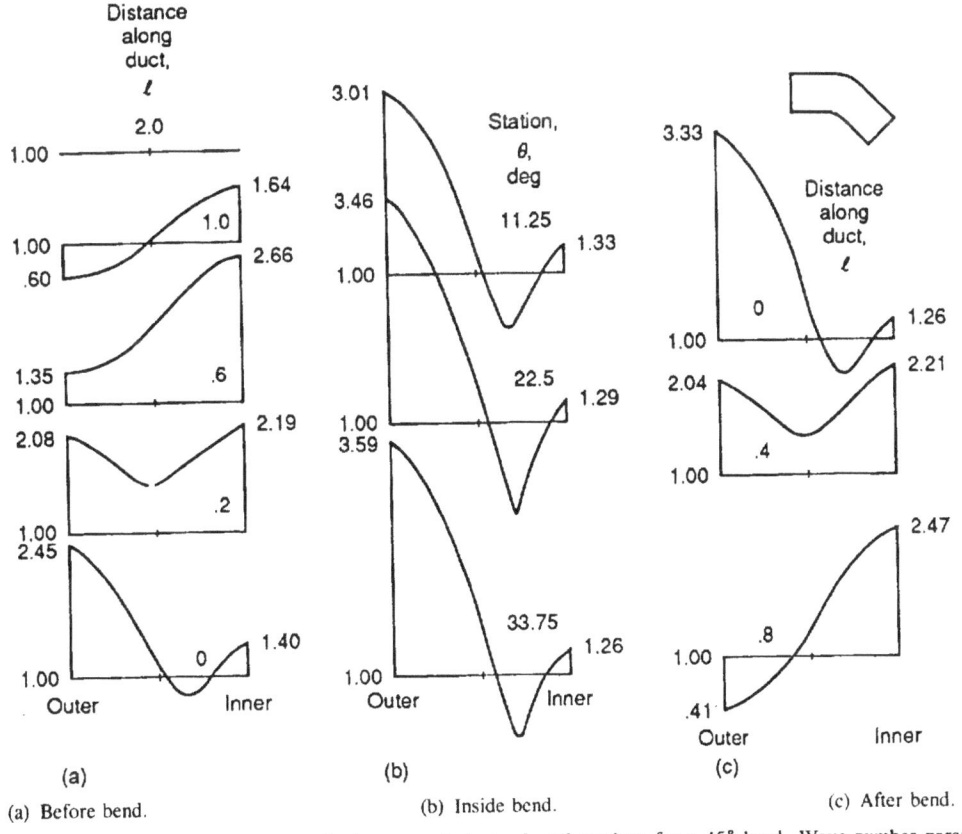

(a) Before bend.

(b) Inside bend.

(c) After bend.

Figure 8.27.—Distribution of pressure amplitude across duct at selected sections for a 45° bend. Wave number parameter, $kR_1(a - 1) = 3.0$; radius ratio, $a = 9$. From Cabelli (1980).

Several other test results are given in figures 8.28 to 8.30. Cabelli's discussion of his findings is given here.

...Figure [8.28] displays the distribution of the pressure amplitude at four sections across the duct. Station 1 was located at the plane of the discontinuity, whereas stations 2, 3 and 4 were located upstream of the discontinuity at dimensionless distances of 0.33, 0.66 and 1.13 respectively. In all the cases there is good agreement between the experimental and the numerical results....

Cummings [1974] pointed out that "the sound pressure pattern in the (0,0) mode in the curved duct section is not uniform across the duct, although it tends towards uniformity for large bend radii." Furthermore, his results indicated that the variation across the duct became progressively more marked for increasing frequencies.... It should be stressed, however, that in the solutions which were obtained by Cummings the effects of evanescent modes were neglected whereas in the solutions which were obtained by the finite difference method they were not. The interaction of these decaying modes with the propagating modes inside the bend resulted in characteristics of the pressure amplitude which were different from those obtained by Cummings and, in particular, which were dependent on the geometry of the bend. Typically, for a 45° bend with an inner radius [parameter] of 0.125 (i.e., a radius ratio of 9:1), the profile of the pressure amplitude followed the prediction made by Cummings, with the departure from uniformity at the centre of the bend increasing with increasing magnitude of the wave number. In contrast, for a 180° bend with the same curvature, the variation from uniformity at the central section was most pronounced for a value of the wave number equal to 2.50 approximately. The distribution of the pressure amplitude at this section, normalized with respect to the value at the outer boundary, is shown in Figure [8.29].... The decay characteristics of evanescent modes in infinite bends have been described by Rostafinski [1976] and his analyis predicted that for any value of the wave number non-propagating modes would be sustained longer in sharper bends. This characteristic is verified in Figure [8.30], which displays the distribution of the pressure amplitude at different radial sections in bends with inner radii [parameters] of 0.125

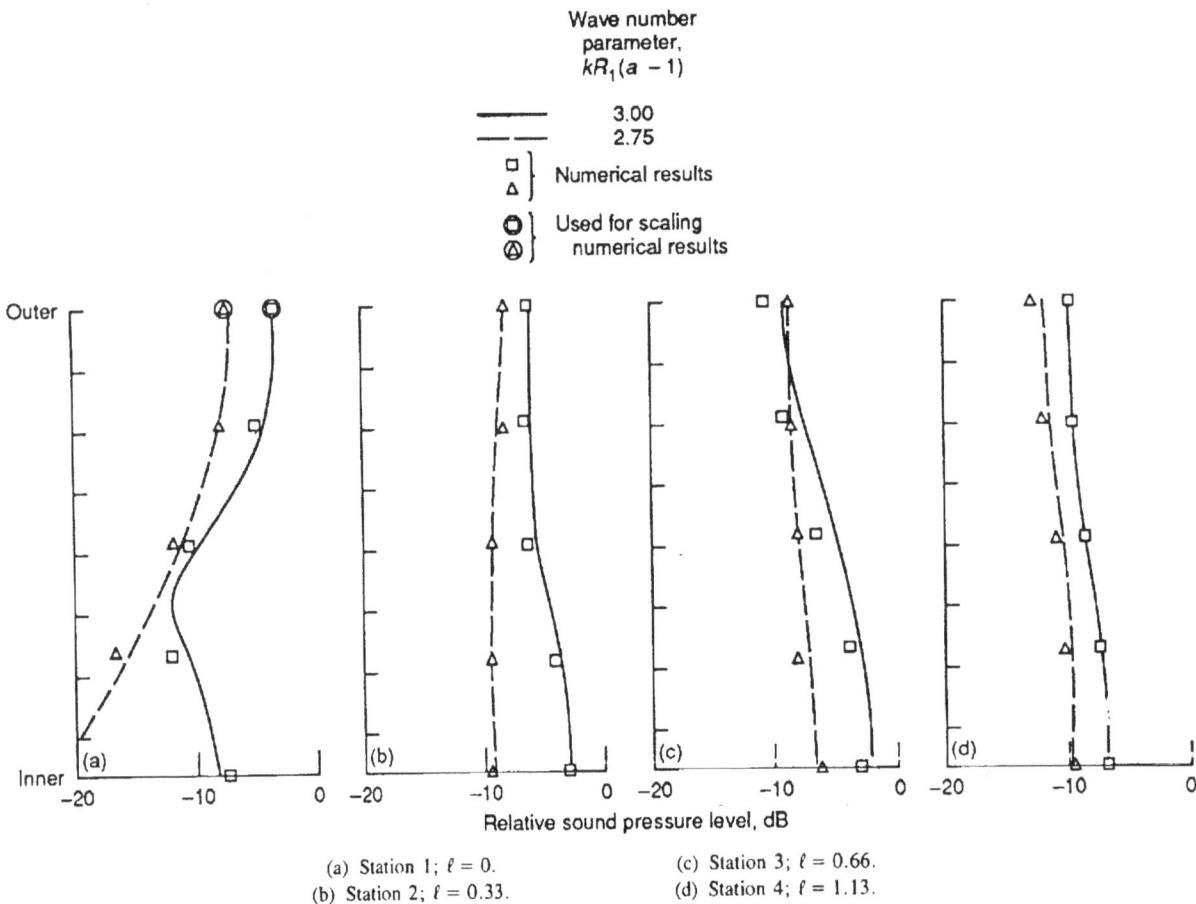

(a) Station 1; $\ell = 0$.
(b) Station 2; $\ell = 0.33$.
(c) Station 3; $\ell = 0.66$.
(d) Station 4; $\ell = 1.13$.

Figure 8.28.—Distribution of pressure amplitude across duct at selected sections upstream of 45° bend. Radius ratio, $a = 9$. From Cabelli (1980).

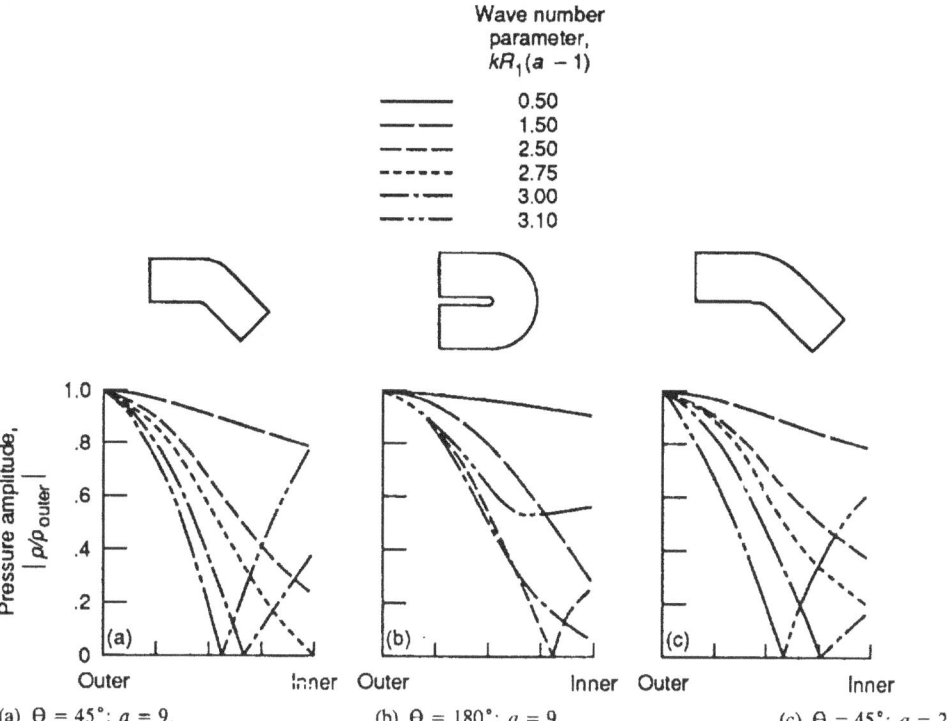

Figure 8.29.—Distribution of pressure amplitude across midsection of bend for various overall bend angles Θ and radius ratios *a*. Numerical results are normalized with respect to outer boundary. From Cabelli (1980).

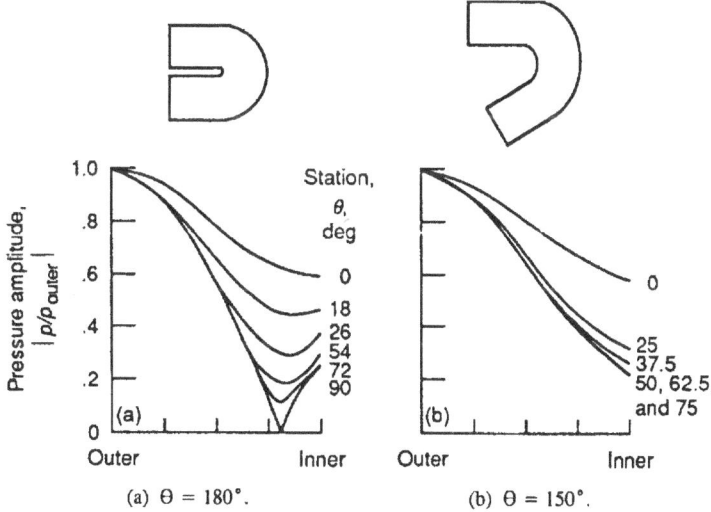

Figure 8.30.—Distribution of pressure amplitude at various sections across bend for two overall bend angles Θ. Wave number parameter, $kR_1(a - 1) = 2.5$. Numerical results are normalized with respect to outer boundary. From Cabelli (1980).

and 0.50 respectively [radius ratios, 9:1 and 3:1] and for a value of the wave number parameter equal to 2.50. The evanescent modes had obviously not decayed in the central section of the sharper 180° bend whereas in the 150° bend with the bigger inner radius, amplitude profiles were identical for angles greater than 50°. In the presence of evanescent modes, the shape of these curves can vary considerably according to the amplitude and the relative phase of each contributing mode. This is again exemplified by the results for the 180° of Figure [8.29] where, with the exception of the solution for $k = 0.5$, evanescent modes were always found to be evident at the plane of symmetry of the bend but were small for k less than 2.0 approximately. The cusps in the distributions of the pressure amplitude of Figures [8.29] and [8.30] were in fact a consequence of the influence of the evanescent modes on the propagating mode. Whilst no experimental pressure measurements were taken radially at the plane of symmetry of the bends, good agreement was obtained with the published experimental results of Cummings for a 180° bend with a radius ratio of 10:1 and a value of the wave number equal to 2.1 approximately.

8.2 Acoustically Lined Bends

Three papers, abstracted here, present data on pressure and particle velocity distribution in curved ducts with acoustically lined walls. Rostafinski (1982) discusses the case of long waves (extremely low frequencies). Several of his data are given in figures 8.31 and 8.32. His comments about those figures are as follows:

> The radial distribution of the particle velocity components and of the acoustic pressure has been analyzed.... Samples of typical profiles in a curved, lined duct are shown in Fig. [8.31]. The particle velocities are nondimensionalized by using v_0. Basically, very slight radial variations have been detected when [conductance] τ and [susceptance] σ were given their full range of values. Also, when the wave was moving down the curved duct, the pressure and axial-velocity profiles remained unchanged except for a gradual decrease in their amplitudes. The profile of the radial components of the particle velocity is different in that its slope changes with distance. The radial velocity distribution exhibits a zero point near the inner wall of the bend. This point is not much displaced toward the center of the duct when a duct three times as wide is used. In Fig. [8.32] these profiles of particle velocity components are compared with distributions calculated for the case of unlined bends of the same geometry and for similar, very low frequencies of acoustic waves. A striking difference exists between the two radial particle velocity distributions. This was to be expected, because a lined duct has a finite value of wall impedance which does not require that the radial particle velocities vanish at the wall. The relative values of the radial velocities shown on the graph have no particular meaning, because the boundary conditions at $\theta = 0$ for the rigid and the lined bends are different.

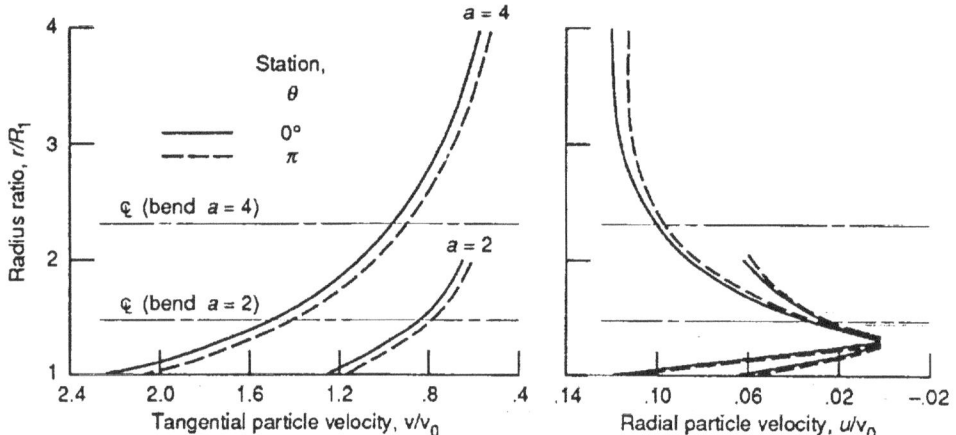

Figure 8.31.—Particle velocity components in lined curved duct for $\sigma = 0.5$, $\tau = 0.1$, and two radius ratios. From Rostafinski (1982).

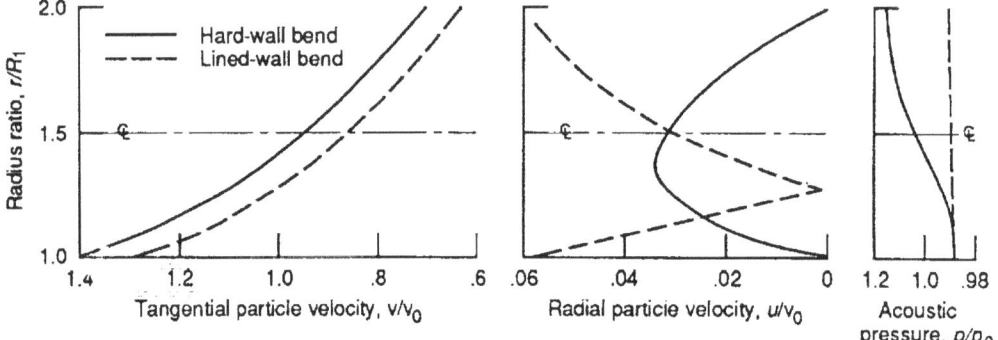

Figure 8.32.—Particle velocities and pressure distribution in lined and unlined curved ducts. Radius ratio, $a = 2$; conductance, $\tau = 0.1$; susceptance, $\sigma = 0.5$. From Rostafinski (1982).

The distribution of the tangential velocity component is not affected by the presence of a lining, except for a generally lower amplitude at all radii.

The radial distribution of the acoustic pressure is markedly changed by the presence of a lining. The data are nondimensionalized by using p_0, the acoustic pressure corresponding to reference particle velocity v_0. Within a hard-walled bend, the pressure is characteristically higher at the outside wall, as was first documented by Cummings [1974]. In a lined, curved duct, it is practically independent of the radial position.

Myers and Mungur (1976) calculated pressure mode shapes for a bend of radius ratio $a = 2$ and a representative frequency corresponding to $kR_1(a - 1) = 6.0$. Their graphs illustrate the capability of their computational code in the area of complex numbers. On two graphs (fig. 8.33) a single type of acoustical lining is considered with nondimensional wall admittance $\eta = 1 + i0.25$, and data are given for the real and imaginary parts of the wave potential function. They add:

It is perhaps worthwhile to note that there is no plane-wave mode in the curved duct, even if all the walls are rigid. The (0,0) mode has a significant radial variation, and this variation becomes sharper the higher the frequency. Thus, for example, a planar piston source at $\theta = 0$ will necessarily excite several modes in the duct unless the frequency is very low [see Rostafinski (1972)]. Implications of this are discussed in the next section in terms of the flow of acoustic power in the duct.

Finally, unusual data have been published by Baumeister (1989). In a set of five graphs (fig. 8.34) he examines the pressure field in an S-shaped duct of five levels of offset h/b_a. The wall lining is of the extended reacting type; in contrast, other authors have considered locally reacting liners, such as perforated plates. The wall properties were taken to be $\epsilon_w = 1.0 - i2.83$ and $\mu_w = 4.1$, where ϵ_w and μ_w are dimensionless property constants. These properties (as stated in the reference) are associated with nearly maximum absorption of a plane pressure wave in a straight duct at the frequency of unity. In addition, the author comments:

The root-mean-square pressure fields inside the duct are illustrated in Fig. [8.34]...As seen in Fig. [8.34(a)] the pressure remains high in the central portion of the duct with grazing contact along the absorbing wall until it reaches the exit with very little attenuation. In contrast, in Fig. [8.34(e)] the pressure field comes in nearly normal contact with the wall and quickly dies out giving rise to the much larger power attenuation....

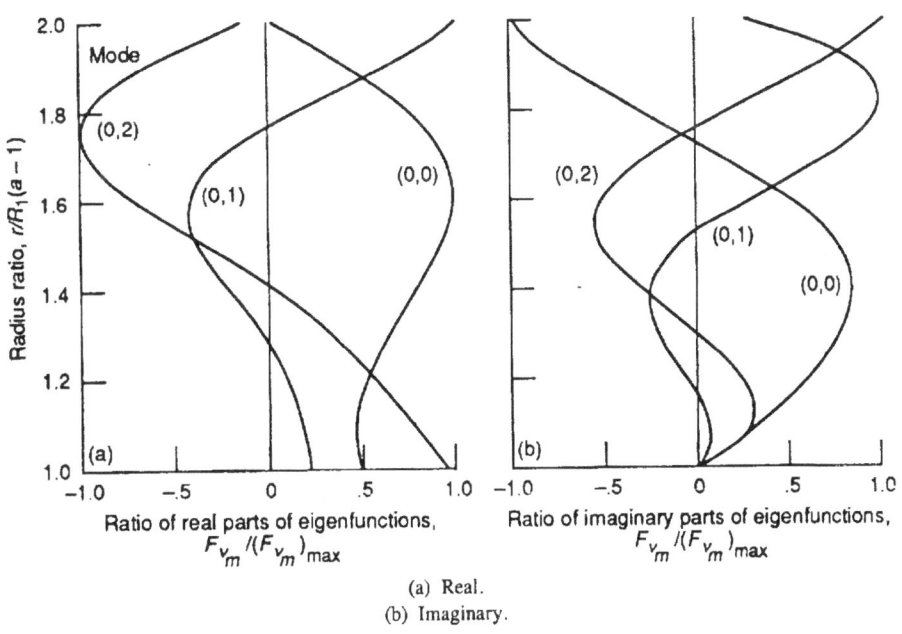

(a) Real.
(b) Imaginary.

Figure 8.33.—Pressure mode shapes for lined duct bend. Nondimensional wall admittance, $\eta = 1 + i\,0.25$; radius ratio, $a = 2$; angular coordinate in cross section, $\alpha = 0.5$; wave number parameter, $kR_1(a - 1) = 6$. From Myers and Mungur (1976).

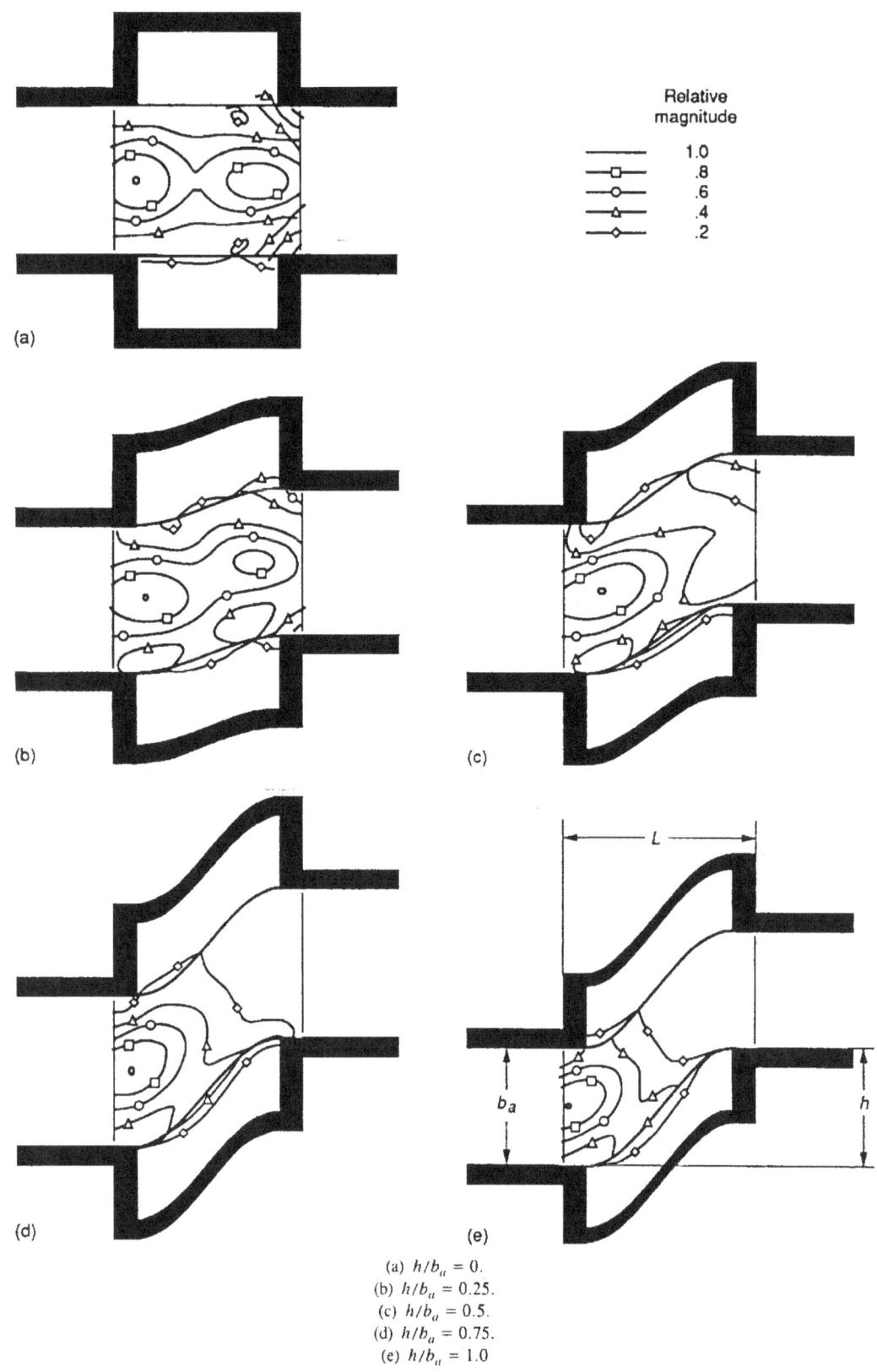

(a) $h/b_a = 0$.
(b) $h/b_a = 0.25$.
(c) $h/b_a = 0.5$.
(d) $h/b_a = 0.75$.
(e) $h/b_a = 1.0$

Figure 8.34.—Effect of absorbing wall offset on pressure field contours with nonreflecting exit for dimensionless length $L = 0.75$ and dimensionless frequency $f = 1$ for various ratios of duct offset height to entrance height h/b_a. From Baumeister (1989).

9.0 Sound Reflection and Transmission

Sound reflection and transmission in bends undoubtedly need further work. Some interesting experimental data have been generated, but the theoretical developments lack general understanding. A degree of confusion (reflections versus evanescent, backward-running waves at discontinuities) can be noticed in various papers.

Although Rostafinski (1972) presents a comprehensive theory on evanescent waves at extremely low frequencies, he makes no reference to sound reflection in bends. The first important experimental information on this subject was given by Cummings (1974). His measurement apparatus is shown in figure 9.1. The following quote is taken from his paper:

> Three types of measurement were made on curved bends. The first of these was the pressure transmission coefficient of the whole bend... with a ρc termination in the form of a polyurethane foam wedge. The second experiment was to measure the specific impedance of the duct system...[with] a rigid closed end. [For details on the two bends evaluated by Cummings, see section 8.0.]....The transmission coefficient, T, was calculated from the standing-wave ratio, SWR:

$$T = \sqrt{1 - \left(\frac{\text{SWR} - 1}{\text{SWR} + 1}\right)^2}.$$

> The impedance was calculated from the usual standing-wave theory.

> The pressure transmission coefficient gave a measure of how much sound was reflected back from the bend down the duct towards the source: in other words, how severe a discontinuity the bend presented to the incident sound wave. The impedance was a useful measure of the effective acoustic path length of the bend, a quantity which, as will be seen, is of major interest in any calculations on curved bends. The radial factor for the sound field gave a measure of how much the sound field in the curved portion of the duct deviated from that for a plane wave in a straight duct....

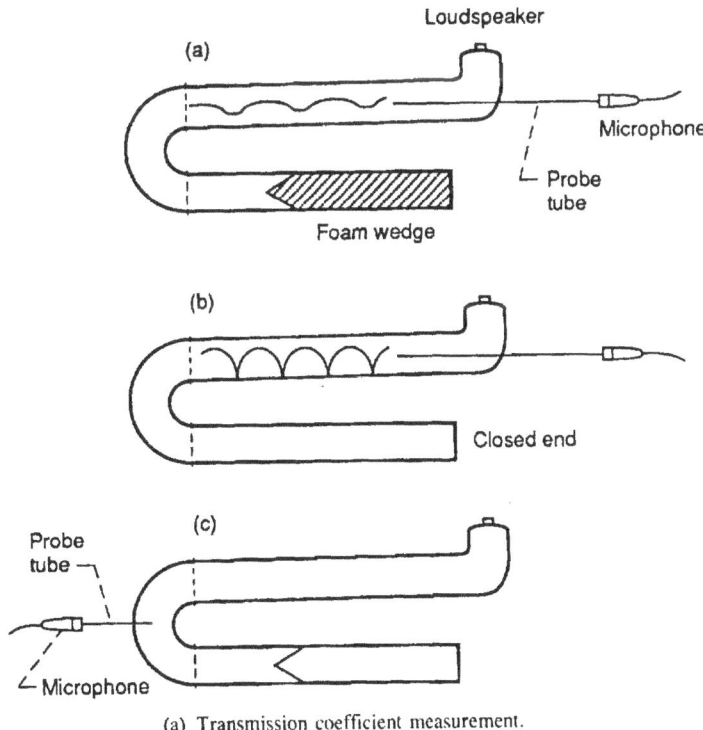

(a) Transmission coefficient measurement.
(b) Impedance measurement.
(c) Radial factor measurement.

Figure 9.1.—Measurement apparatus. From Cummings (1974).

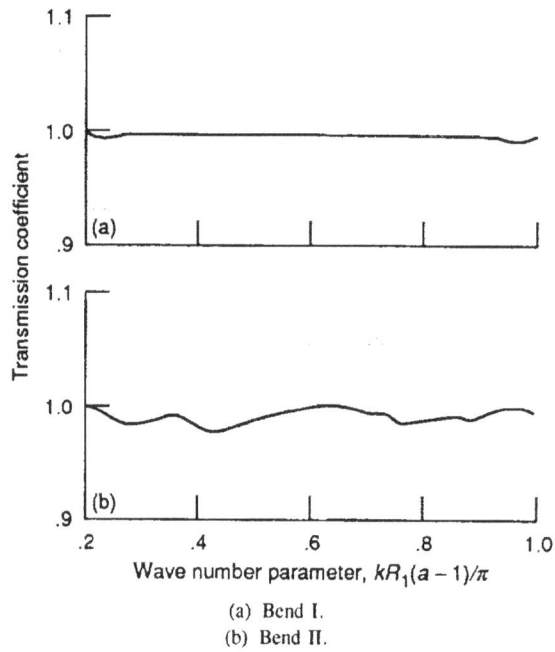

(a) Bend I.
(b) Bend II.

Figure 9.2.—Transmission coefficients. From Cummings (1974).

The results of the pressure transmission coefficient measurements for bends I and II are shown in Figures [9.2] (a) and (b). The values of T are plotted against the parameter $[(R_2 - R_1)k/\pi = bk/\pi]$ (this is equal to 1 at the cut-on frequency of the first higher-order mode in duct sections 1 and 3.) It is seen that for bend I (of moderate sharpness), T is practically 100% over the whole of the frequency range.... For bend II (of severe sharpness) T is still very close to 1, and does not fall below 97.5%. It is felt that bend II represents the sharpest bend likely to be encountered in practice—thus, for all practical purposes, the transmission coefficient of all curved elbow bends, for $[(R_2 - R_1)k/\pi] < 1$, may be considered to be unity.

In his conclusions, Cummings restates that it has been shown that curved duct bends, even those of severe sharpness (e.g., $a = 10.3$), transmit sound extremely efficiently and that higher-order mode generation at bends is of secondary importance. Recall that Cummings restricts most of his remarks to the basic mode. He says, "Sound transmission through rectangular and circular section curved elbow bends has been discussed, with particular reference to the frequency region where only the (0,0) mode propagates."

A completely different picture emerges from the work of Cabelli (1980). His data on the distribution of pressure amplitudes across sections of a straight duct–bend–straight duct system were discussed earlier. His tests were conducted for wave numbers of 0.5 to 3.0 (not really high frequencies) and in ducts of several total angles but all of radius ratio $a = 9.0$, that is, extremely sharp bends in which serious distortion of particle velocities and pressure distribution profiles should be expected. Figures 9.3 and 9.4 are examples of his results. They also illustrate a good experimental check of his numerical calculations.

Cabelli (1980) devoted a section of his paper to evanescent waves. His conclusions are as follows:

An interesting feature of the propagation of sound in this duct system is concerned with the generation of higher order modes at the discontinuity. Figure [8.26] indicated that for values of less than π, the first (evanescent) mode was the most significant among the higher orders....

The magnitude of the higher order modes was also found to be affected by the angle of the bend. This, however, was not a simple relationship and any trend observed for a range of values of Θ was only valid at the particular value of the wave number parameter. It is significant, however, that the greatest amplitude of the first cross mode which was found at (say) the upstream discontinuity was generally not for the bend with the greatest angle. Furthermore, as the value of k was increased, the angle of the bend for which this component was greatest decreased. Consider for instance bends with an inner radius of 0.125 units $[a = 9.0]$. At a value of k of 2.5, the normalized magnitude of the first cross mode reached a maximum of 0.7 when the angle of the bend was approximately

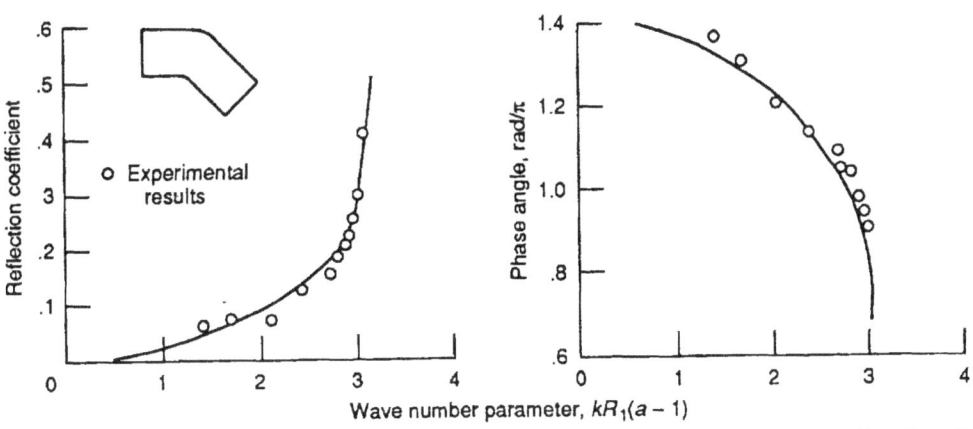

Figure 9.3.—Variation of reflection coefficient and phase angle with wave number. Overall bend angle. $\Theta = 45°$; radius ratio, $a = 9$. From Cabelli (1980).

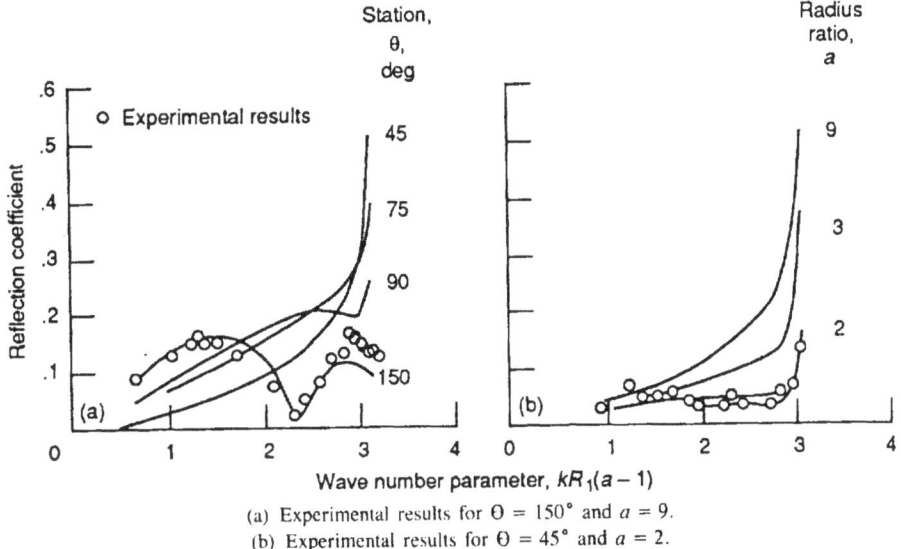

(a) Experimental results for $\Theta = 150°$ and $a = 9$.
(b) Experimental results for $\Theta = 45°$ and $a = 2$.

Figure 9.4.—Variation of reflection coefficient with wave number, radius ratio a, and overall bend angle Θ. From Cabelli (1980).

equal to 90°. When k was increased to 3.1, a maximum in excess of 2 was reached for a bend angle of 30° approximately. The corresponding values for the normalized magnitude of the second order mode were 0.10 and 0.11 approximately. The significance of this result is that any solution which relies on matching boundary conditions at the discontinuity and which is satisfactory for a bend with a given turning angle will not necessarily be satisfactory for a bend with a smaller turning angle unless the number of terms used to satisfy continuity of pressure and velocity is made correspondingly higher.

El-Raheb and Wagner (1980), in a study of resonances in duct corners and bends, commented on reflections but without evaluating them. Figure 9.5 gives pressure distribution in a duct corner (miter bend) and in a similar circular bend, in both cases for frequencies corresponding to the first resonance (1,0). For the corner the first resonance happens at 1306 Hz, and in a bend it happens at 1373 Hz. They comment:

At frequencies lower than $\omega^*(1,0)$, the bend's isobars are lines radiating from its center of curvature. As the frequency increases, constant pressure lobes form near midlength along the larger circumference suggesting an increase in transverse wave activity. These standing waves are generated as a consequence of reflections from the duct's curved boundary. The reflections produced by a corner are more intense than those initiated by the smoothly varying boundary of a bend. Note the difference in transverse pressure gradients (across lobes) between bend and corner for $\omega^*(1,0)$. As a result,

43

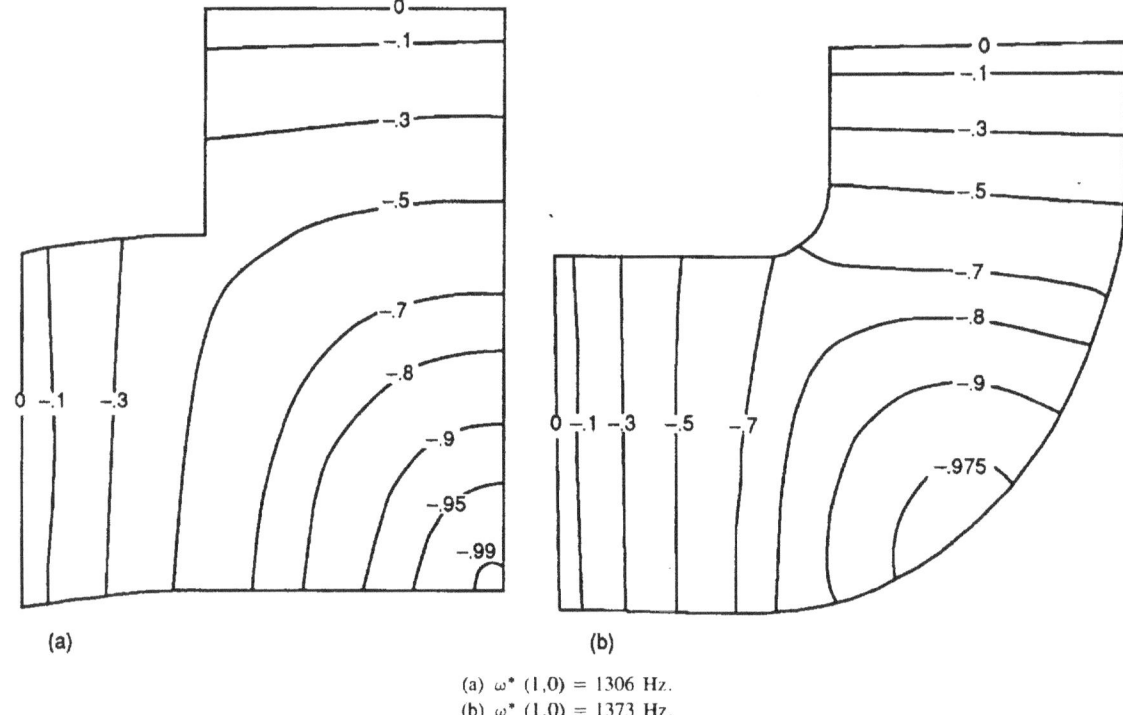

(a) $\omega^*\,(1,0) = 1306$ Hz.
(b) $\omega^*\,(1,0) = 1373$ Hz.

Figure 9.5.—Isobars at first resonant frequencies $\omega^*\,(1,0)$. From El-Raheb and Wagner (1980).

the longitudinal phase velocity and longitudinal pressure gradient are lower in the corner than in the bend since in the latter less energy is consumed in transverse excitations. Consequently, $\omega^*(1,0)$ in the corner is lower than in the bend.

Keefe and Benade (1983) did not study reflections in bends for the simple reason that reflections are imperceptible for the long waves they considered. (This is true even in the 90° miter corner studied by Lippert, 1954.) Nevertheless, they considered the evanescent waves and concluded that for low frequencies

...the effect of the evanescent modes is very small compared to the propagating mode, since the discontinuity impedance is much less in magnitude than the propagating mode impedance.... The assumption made that the evanescent modes response is negligible relative to that of propagating mode is therefore shown to be valid.

Tam (1976), who in his applied mathematics paper (using a procedure based on the Galerkin method) developed a method for determining angular wave numbers, makes the following interesting comments:

As frequency increases there are more and more propagating modes just as in the case of a straight duct. It is to be noted that even at moderately large values of ω the angular wave numbers [ν] of the propagating modes are quite large. This implies (verified by actual computation) that the phases of the transmission and reflection coefficients will oscillate very rapidly between positive and negative values even with small changes in [total bend angle] or ω. Thus in considering the transmission of sound through a duct system good accuracy must be retained in joining the solutions together at the two ends of a curved bend so as to ensure correct matching of phases.

Discussing the structure of his transmission matrix, Tam states that the physical implication of the developed diagonal symmetry is that the transmission coefficient of the mth mode due to an incident wave of the nth mode is the same as the transmission coefficient of the nth mode due to an incident wave of the mth mode. This interesting property does not seem to have been noticed before.

Osborne (1976) indicates that analysis of a 202.5° bend in the fundamental mode, for which the angular wave number is 4.25 (the frequency being 87 percent of the first propagating cross-mode frequency), shows no variation in sound pressure distribution between angular sections spaced 22.5° apart. The experimental evidence, therefore, suggests that a bend of 45° when excited in the first few modes does not constitute an appreciable discontinuity and that reflections are negligible. Further, he notes that the contribution to the sound field by evanescent modes is more difficult to determine. It might be expected that for bends of large radius ratio, and particularly in the fundamental mode, the effect of nonpropagating modes will play a large part in establishing the transition of the sound field between the straight duct and the bend. In Osborne's (1976) work it was difficult to discern any effect that could be attributed specifically to those modes.

Finally, in a study of sound propagation in an acoustically lined S-shaped bend Baumeister (1989) touches upon the subject of evanescent and reflected waves. It is discussed here in section 11.0.

The important and useful theory of reflections finds its application in circular bends equipped with axial turning vanes, partitions that effectively create two bends from one. This special study is presented in section 13.0.

10.0 Impedance of Bends and Resonances

Cummings (1974) first calculated and experimentally verified the impedance of curved ducts. He describes his procedure and his interesting developments and data as follows:

> Figures [10.1 and 10.2] show measured and predicted data on the reactance ratio of bends I and II [both 180°; for bend I, $a = 1.587$; for bend II, $a = 10.3$], for various frequencies, with section 3 terminated in a rigid wall. It was thought that possibly it may be sufficient to neglect completely higher-order mode generation and pressure and velocity non-uniformity at discontinuities...and simply to correct the length of the bend at the centreline for the modified acoustic propagation velocity (along the centreline) in the curved section (the circumferential wavenumber, k_c, is equal to $[v/R_m]$ where $[R_m]$ is the mean radius of the bend). Thus the impedance (assumed to be entirely reactive) is given by
>
> $$\zeta_B = i \cot \left[k \left(\Theta R_m \frac{k_c}{k} + \ell \right) \right].$$
>
> The angular wavenumber $[v]$ was [calculated]. The reactance curves...are plotted out in Figures [10.1 and 10.2]. In the absence of appropriate data, one would intuitively take the bend's effective length to be equal to the length ℓ_m along the centreline, with impedance given by
>
> $$\zeta_B = i \cot \left[k (\ell_m + \ell) \right]$$
>
> [where ℓ is the length of a straight duct downstream from the bend].
>
> The reactance corresponding to this expression is also plotted out in Figures [10.1 and 10.2]. It is seen that in all cases, merely correcting the median length of the bend gives predictions in good agreement with measurements. The impedance predicted when using the bend's uncorrected median length is (except at low frequencies) not in agreement with experiment. In Figure [10.2] the agreement between the corrected median length of the bend and experiment falls off somewhat towards the anti-resonance at about $[k(R_2 - R_1)/\pi] = 0.14$. Accordingly, the reactance predicted...(with no higher-order modes taken into account...)is also plotted out. This is in good agreement with the experimental results. Predictions were not made in which one or more higher order modes were used....
>
> It is apparent that simply correcting the bend's median length produces satisfactory results in most cases. If more accurate predictions are required, then the zero-order solution (whose associated computing cost is reasonable) gives sufficiently accurate results.
>
> From this it is surmised that the principal effect of the curved bend section is not generation of higher-order modes and associated reflection of the acoustic wave from the bend, but a change in the propagation characteristics in the curved section. Certainly, higher-order modes are generated at each discontinuity, but their effects on the bend's overall behaviour seem slight.

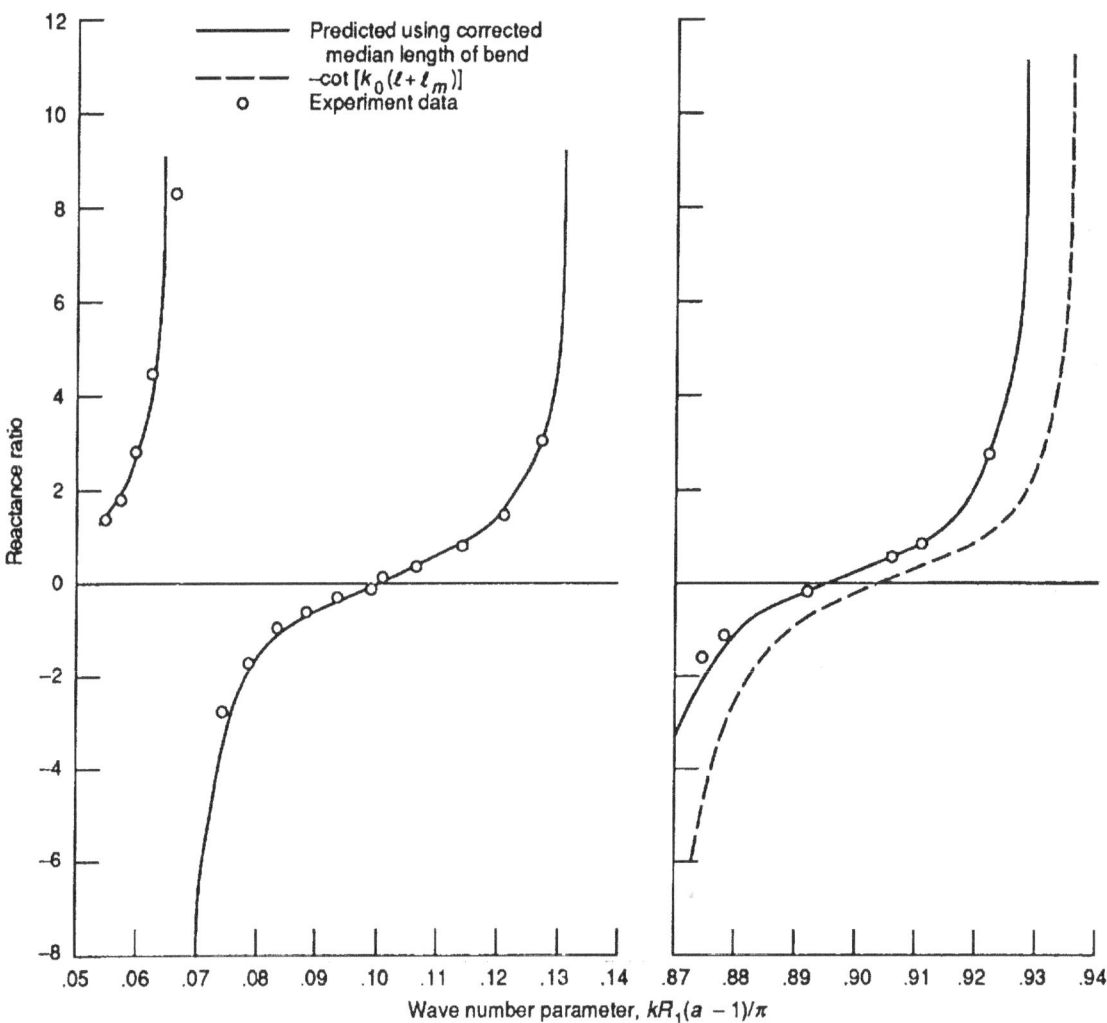

Figure 10.1.—Reactance ratios for bend I. From Cummings (1974).

Rostafinski (1974b; 1976) gave an analytical expression for the impedance of bends and based it on the length of the bend centerline. Keefe and Benade (1983) noted that Rostafinski's results are incorrect because his work predicts an increase in the wave impedance of a bend relative to a straight pipe at low frequencies, whereas a decrease is always observed. They indicated that the impedance Z of a bend is equal to the impedance $Z_0 = \rho c$ of a straight pipe multiplied by ν/kR_m. They also write, for future reference, that the wave admittance Y equals $Y_0(kR_m/\nu)$, where $Y_0 = 1/Z_0$. They draw the most interesting conclusion that the shifts in wave admittance and phase velocity in a curved rectangular duct, relative to a straight duct, are equal even if the small radial variations of the pressure are included. Effectively, their expression for long-wave phase velocity is $c(kR_m/\nu)$.

Recalculating Rostafinski's expression for wave impedance in bends (long waves; (0,0) mode only) based on the bend radius where the phase velocity equals v_0 (piston's $v_0 e^{i\omega t}$, see figs. 1.1(a) and (b)) yields $Z = Z_0(kR_1/\nu)[(a - 1)/\ln a]$. Keefe and Benade's equation and Rostafinski's corrected equation yield identical results, some of which are given in figure 10.3.

Keefe and Benade (1983) called attention to the often overlooked fact that the natural frequencies of the air column in curved pipes

are affected not only by changes in the impedance and velocity produced by pipe curvature, but also by discontinuity effects that arise at the junctions between pipe segments of differing curvature. Any such junction discontinuity may be represented to good approximation by a series inertance plus a small resistive term which does not affect the resonance frequency although it may contribute

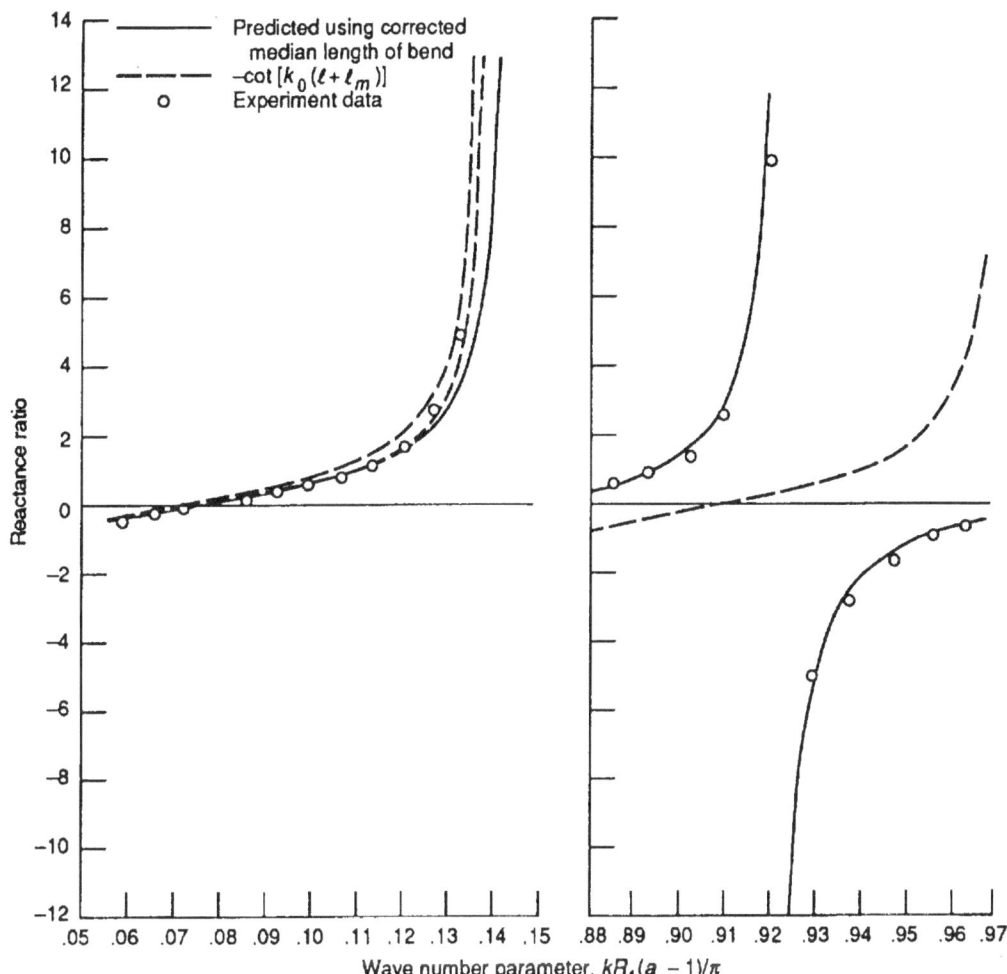

Figure 10.2.—Reactance ratios for bend II. From Cummings (1974).

significantly to the overall damping of the modes. This is because there is a local change in the flow field at the junction, but pressure is continuous. The resulting perturbation in the local kinetic energy density is represented in the transmission line by a complex inertance in series.

El-Raheb and Wagner (1980) conducted an extensive (numerical) analysis of wave propagation in miter joints, bends, and three-way branches. The bend considered consisted of a 90° elbow ($a = 5$) with a nearly square cross section. Calculations involved three methods: the Green function's formulation, the finite difference method, and an analytical approach (eigenfunction expansion). Their results are summarized in figure 10.4, which compares average loading in a 90° bend for the three methods. The agreement is excellent indeed and the results are impressive, resonant frequencies having been calculated for a wide frequency range. In table 10.1 an interesting comparison is made; it shows that, except for one case, the bend lowers the resonant frequencies relative to those in a straight duct.

El-Raheb (1980) evaluated resonant frequencies in a network of rigid ducts with four bends. Using eigenfunction expansion he first calculated eigenvalues (angular wave numbers, see fig. 6.2) and then evaluated the load factor for the cases given in table 10.2. Additional cases included the addition of straight ducts downstream from the bend (cases C1 and E1). The results of this extensive analysis are given in figures 10.5 to 10.7 and table 10.3. El-Raheb's conclusions are as follows:

The basic qualitative aspects of the acoustic wave propagation in rigid finite length networks involving straight and curved ducts are summarized with emphasis on the comparison with the equivalent one-dimensional behavior:

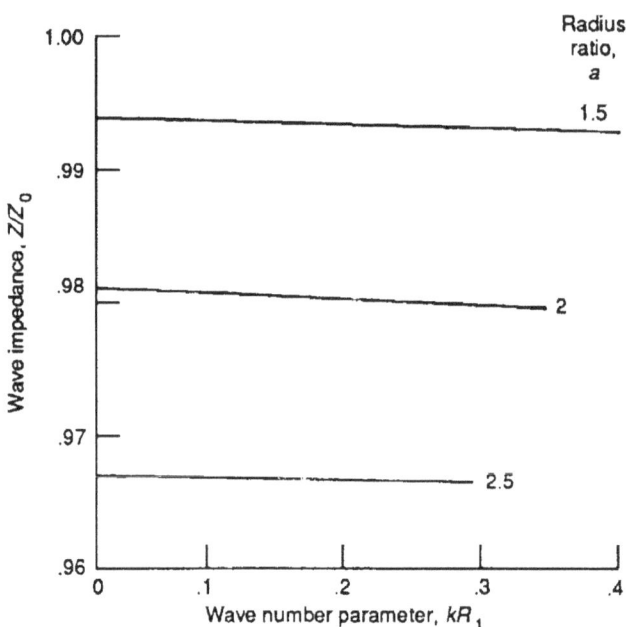

Figure 10.3.—Wave impedance of long waves in bends.

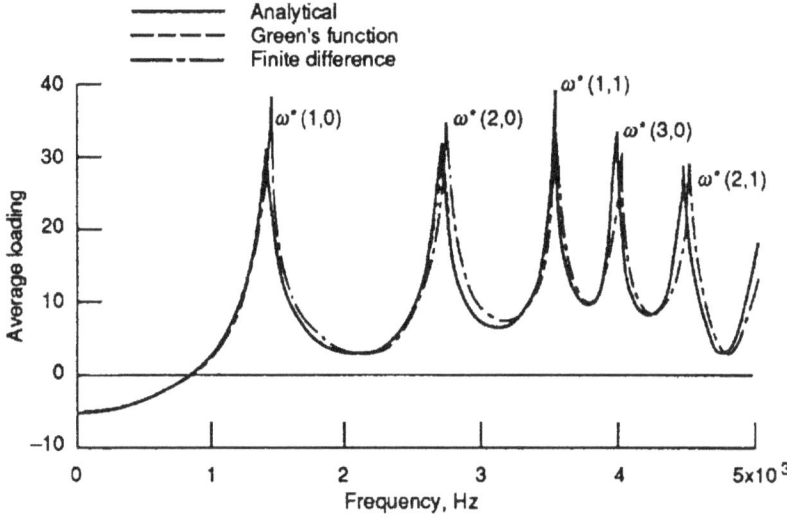

Figure 10.4.—Average loading for 90° circular bend by three calculation methods. From El-Raheb and Wagner (1980).

TABLE 10.1.—ACOUSTIC RESONANCES

[From El-Raheb and Wagner (1980).]

Type (m,n)	Two-dimensional straight	90° bend	
		Analytical	Green's function (76 elements)
	Resonant frequency, Hz		
(1,0)	1388.9	1376.2	1373.2
(2,0)	2777.8	2669.4	2670.1
(1,1)	3419.7	3503.1	3505.8
(3,0)	4166.7	3957.5	3961.3
(2,1)	4181.1	4463.6	4461.5
(3,1)	5208.3	------	------
(4,0)	5555.6	5040.8	5039.5

TABLE 10.2.—GEOMETRY OF CASES B–E

[From El-Raheb (1980).]

Case	Convex (inner), R_1	Concave (outer), R_2	Radius ratio, a
	Radii of bend wall, m		
B	0.0508	0.1524	3
C	.0508	.254	5
D	.1016	.3048	3
E	.1524	.4572	3

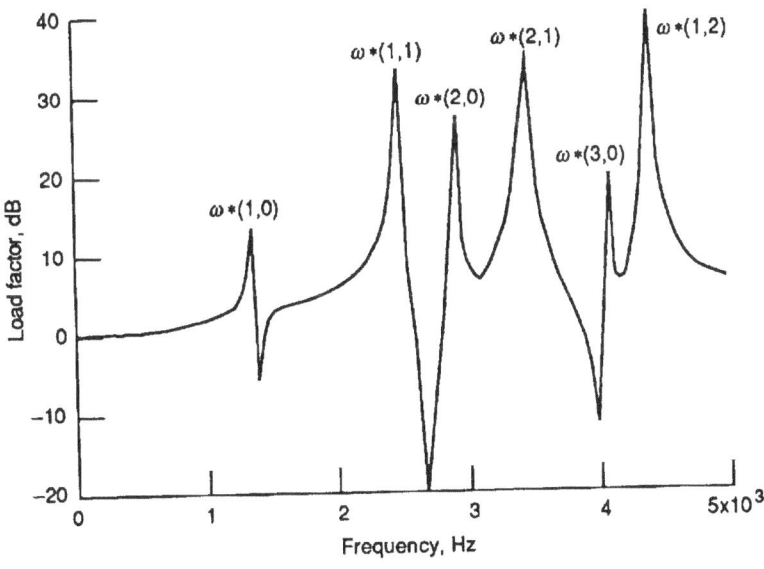

Figure 10.5.—Variation of load factor with frequency for case E from table 10.2. Radii of convex (inner) and concave (outer) walls of bend, respectively: $R_1 = 0.1524$ m; $R_2 = 0.4572$ m. Radius ratio, $a = 3$; speed of sound, $c = 1270$ m/s. From El-Raheb (1980).

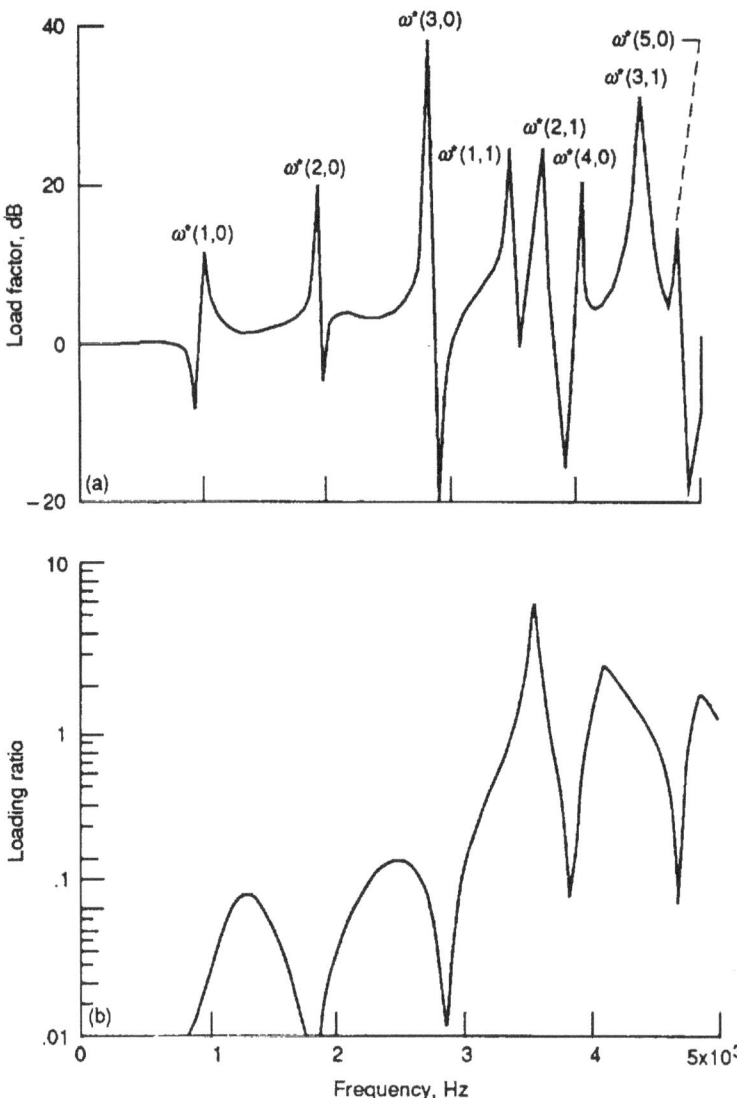

(a) Load factor. Radii of convex (inner) and concave (outer) walls of bend, respectively: $R_1 = 0.1524$ m; $R_2 = 0.4572$ m.

(b) Loading ratio. $R_1 = 0.0508$ m; $R_2 = 0.254$ m.

Figure 10.6.—Variation of load factor and loading ratio with frequency for case C_1 from table 10.3. Centerline length of straight duct, $\ell_s = 0.4064$ m; radius ratio, $a = 5$; speed of sound, $c = 1270$ m/s. From El-Raheb (1980).

(1) The duct curvature introduces a shift in the longitudinal acoustic resonances of the system. The shift fluctuates from positive at large wavelengths to negative for $\omega^*(1,1) > \omega^* > 1/2\ \omega^*(1,1)$ back to positive beyond $\omega^*(1,1)$ following a cyclic and involved mechanism which depends on the position of ω^* relative to $\omega^*(1,n)$ and on wavelength [where n is the transverse acoustic mode shape number]. The magnitude of the shift increases uniformly with (ℓ_b/ℓ_s).

(2) The acoustic loading on the bend grows uniformly with duct width as a result of radial vibrations initiated by curvature. The loading intensity rises continuously as the fundamental radial resonance is approached beyond which the rate drops slightly. The loading is independent of bend curvature although λ is implicitly responsible for triggering the overpressure.

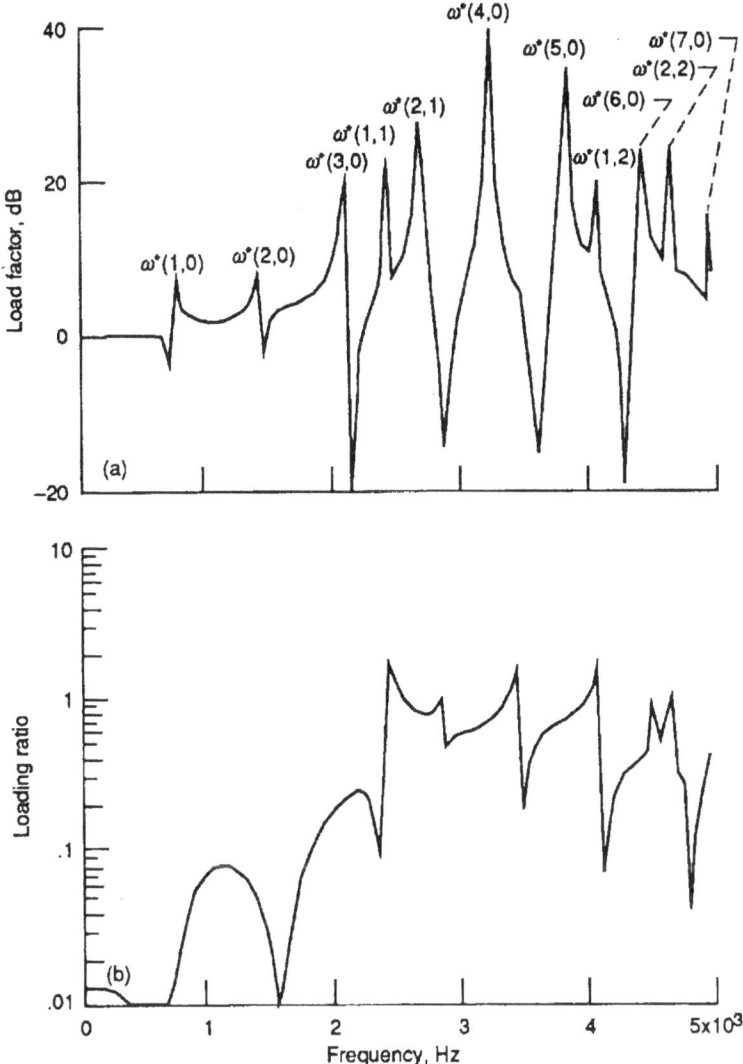

(a) Load factor. Radii of convex (inner) and concave (outer) walls of bend, respectively: $R_1 = 0.0508$ m; $R_2 = 0.254$ m. Radius ratio, $a = 5$.

(b) Loading ratio. $R_1 = 0.1524$ m; $R_2 = 0.4572$ m. Radius ratio, $a = 3$.

Figure 10.7.—Variation of load factor and loading ratio with frequency for case E_1 from table 10.3. Centerline length of straight duct, $\ell_s = 0.4064$ m; speed of sound, $c = 1270$ m/s. From El-Raheb (1980).

(3) The asymmetry in acoustic pressure, resulting from bend curvature, attenuates rapidly as it propagates along a straight duct beyond a bend interface and reaches a near uniform distribution after a short transition distance, so long as $\omega^* < \omega^*_{cof}$. Beyond the first cutoff frequency, the standing radial waves propagate resulting in pressure asymmetry, and loading comparable to the reactive loading on the bend.

(4) Longitudinal system resonances often relocate away from transverse resonances to produce a more even separation and satisfy minimum energy requirements.

TABLE 10.3.—ACOUSTIC RESONANCES FOR
CASES C_1 AND E_1

[From El-Raheb (1980).]

Case	Type	m,n	Nondimensional angular frequency for three-dimensional duct, ω^*		Difference between equivalent straight duct and bend resonances, $100\left(1-\dfrac{\omega^{*e}_{3D}}{\omega^*_{3D}}\right)$, percent
			With bend	Straight	
C_1	Longitudinal	1,0	1015.8	983.3	3.2
	Longitudinal	2,0	1952.3	1966.6	−.7
	Longitudinal	3,0	2848.4	2949.9	−3.6
	Cutoff	0,1	------	3125.0	---
	Radial	1,1	3510.6	3276.1	6.7
	Radial	2,1	3734.4	3692.3	1.1
	Longitudinal	4,0	4063.4	3933.2	3.2
	Radial	3,1	4515.4	4297.4	4.8
	Longitudinal	5,0	4818.7	4916.5	−2.0
E_1	Longitudinal	1,0	730.7	717.4	1.8
	Longitudinal	2,0	1420.5	1434.7	−1.0
	Longitudinal	3,0	2073.5	2152.1	−3.8
	Cutoff	0,1	------	2083.3	---
	Radial	1,1	2376.0	2203.4	7.6
	Radial	2,1	2662.2	2529.6	5.0
	Radial	3,1	------	2995.3	---
	Longitudinal	4,0	3200.3	2869.5	11.5
	Radial	4,1	------	3546.0	---
	Longitudinal	5,0	3807.4	3586.9	6.1
	Radial	5,1	------	4147.9	---
	Cutoff	0,2	------	4166.7	---
	Radial	1,2	4050.0	4228.0	4.0
	Longitudinal	6,0	4418.4	4304.2	2.6
	Radial	2,2	4654.5	4406.8	5.3
	Radial	3,2	------	4689.6	---
	Radial	6,1	------	4781.9	---
	Longitudinal	7,0	4998.2	5021.6	−.5

11.0 Attenuation of Sound in Bends

Attenuation of propagating modes in hard-wall bends may result, almost exclusively, from backrunning waves reflected from the cylindrical boundaries. These losses have been examined and test checked by Cummings (1974) and by Cabelli (1980). This subject is covered in section 9.0.

A different source of attenuation in bends has been evaluated by Keefe and Benade (1983): viscous energy dissipation in curved ducts. Their remarks and conclusions are as follows:

> A wave propagating in a curved duct is modified by the presence of viscosity. There are viscous (and thermal) losses at the walls, viscous bulk losses in the interior of the fluid, and possibly nonlinear viscous losses in the form of time-independent acoustical streaming. The bulk losses in a propagating wave in a straight duct are negligible compared to the wall losses at low to moderate audio frequencies. We show in this section that the bulk losses for a propagating wave in a curved duct of similar geometry to those used in our experiments are orders of magnitude larger than the bulk losses in a straight duct. However, the curved duct bulk losses do not exceed the wall loss in a straight duct of similar cross section....

We compare the bulk loss in the curved pipe to that of the straight pipe by forming the ratio g_b [defined as]

$$g_b \equiv \langle \dot{E}_c \rangle / \langle \dot{E}_s \rangle \sim \frac{1}{2(kR_m)^2} \cdot \left(\frac{a^2 - 1}{a \ln a} \right)^2$$

[where $\langle \dot{E}_c \rangle$ is the time-averaged power loss in a curved duct and $\langle \dot{E}_s \rangle$ is the time-averaged viscous energy loss for a plane wave in a straight pipe].

For the tubing sizes and acoustical frequency used in our experiments g_b turns out to be about 4000. The additional shearing losses in the bulk of the medium in the curved duct are sizable, due to the extra shear present as the propagating wave "turns the corner" of the bend.... Anytime there is curvature in an acoustical waveguide with air as the medium, the shear losses in the bulk of the fluid may become significant relative to the wall loss, even at low frequencies. This is especially true in musical instruments, where the player is extremely sensitive to changes in the damping.... The phase velocity and wave admittance are both increased in a curved duct relative to their straight duct values. Our experiments show that the wave-admittance shift is larger than the phase-velocity shift, although all theories predict that they should be equal. The observed shifts are smaller than the predicted shifts, which suggests that large shearing losses in the bulk of the fluid in the curved duct may be present.

Substantial research has been done on attenuation in cylindrical bends with acoustical linings on the two curved walls and even on all four walls. Sound attenuation is defined as a decrease in mean acoustic energy between the inlet and the outlet sections of a duct segment. The total sound attenuation, also called attenuation of transmitted acoustic power, is measured in decibels.

Grigor'yan (1970) first analyzed sound propagation in a bent duct with its curved walls lined with sound-absorbing material. Using numerical-analytical techniques he evaluated ducts with various degrees of sharpness lined on the inner, the outer, or both curved walls with two different sound-absorbing materials. He concluded that sound-absorbing materials are not equally "sensitive" to the curvature of the bend, that attenuation increases with the sharpness of the bend, and that when only the inner curved wall is lined, attenuation decreases with the sharpness of the bend. In other words, by no means does curvature of a lined duct always substantially increase attenuation of the fundamental mode. The effect depends on the properties of the sound-absorbing material, the lining arrangement (which wall is lined), and the frequency.

Rostafinski (1982) did some studies in the extremely long wave region, compared sound attenuation in straight and curved lined ducts, and checked particle velocity (both tangential and radial) for two values of conductance and susceptance. He calculated sound propagation in a straight duct by using an expression given by Rice (1975)

$$p = \cos \left(\frac{2gy}{H} \right) e^{[i\omega t - (\omega/c)\zeta x]}$$

where the g's are complex roots, $\zeta = \alpha + i\beta$ is a complex wave number with its propagation and attenuation terms, $H/2$ is the half-width of the duct, and x and y are axial and transverse coordinates with $y = 0$ at the duct centerline. Rostafinski calculated sound propagation in bends by classical expansion of the Bessel functions of complex order. The results of his analysis indicate that at extremely low frequencies and for the duct and lining parameters taken into consideration, sound attenuation is less pronounced in bends than in a straight duct of the same wall impedance.

Myers and Mungur (1976) calculated transmission loss in lined straight ducts and lined curved ducts ($a = 2$) for one type of lining but for several bend angles and for a range of wave number parameters $kR_1(a - 1)$. Commenting on their results (figs. 11.1 and 11.2), they noted that a curved duct generally seems to attenuate the sound field more than a straight duct, the difference being more pronounced at the higher frequencies. At $kR_1(a - 1) = 4$ the transmission losses in the curved and straight sections are nearly equal; the curves are coincident in figure 11.2(a). The difference increases at the higher wave number parameters.

An important study by Ko and Ho (1977) on sound attenuation in acoustically lined curved ducts in the absence of fluid flow is based on numerical evaluation of equations obtained by the separation of variables.

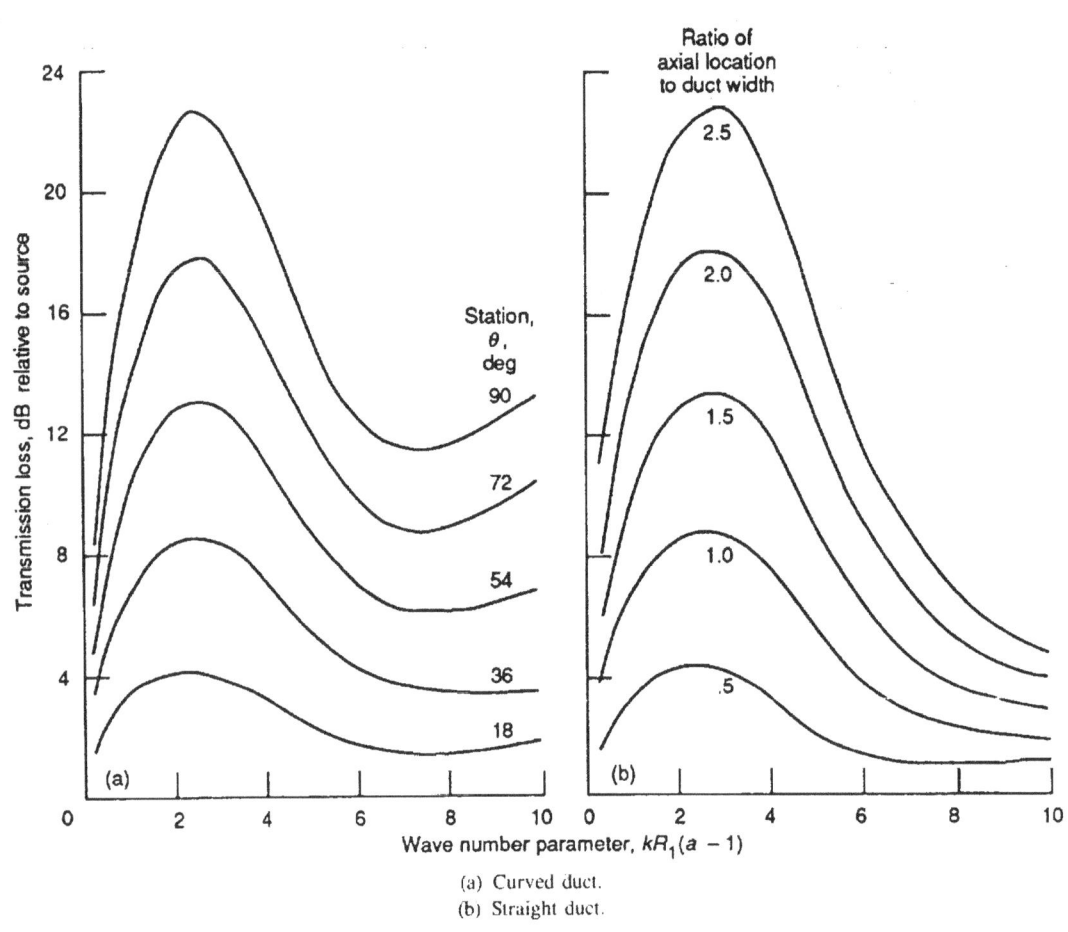

(a) Curved duct.
(b) Straight duct.

Figure 11.1.—Transmission loss in curved and straight ducts. Radius ratio, $a = 2$; nondimensional wall admittance, $\eta = 1 + i0.25$. From Myers and Mungur (1976).

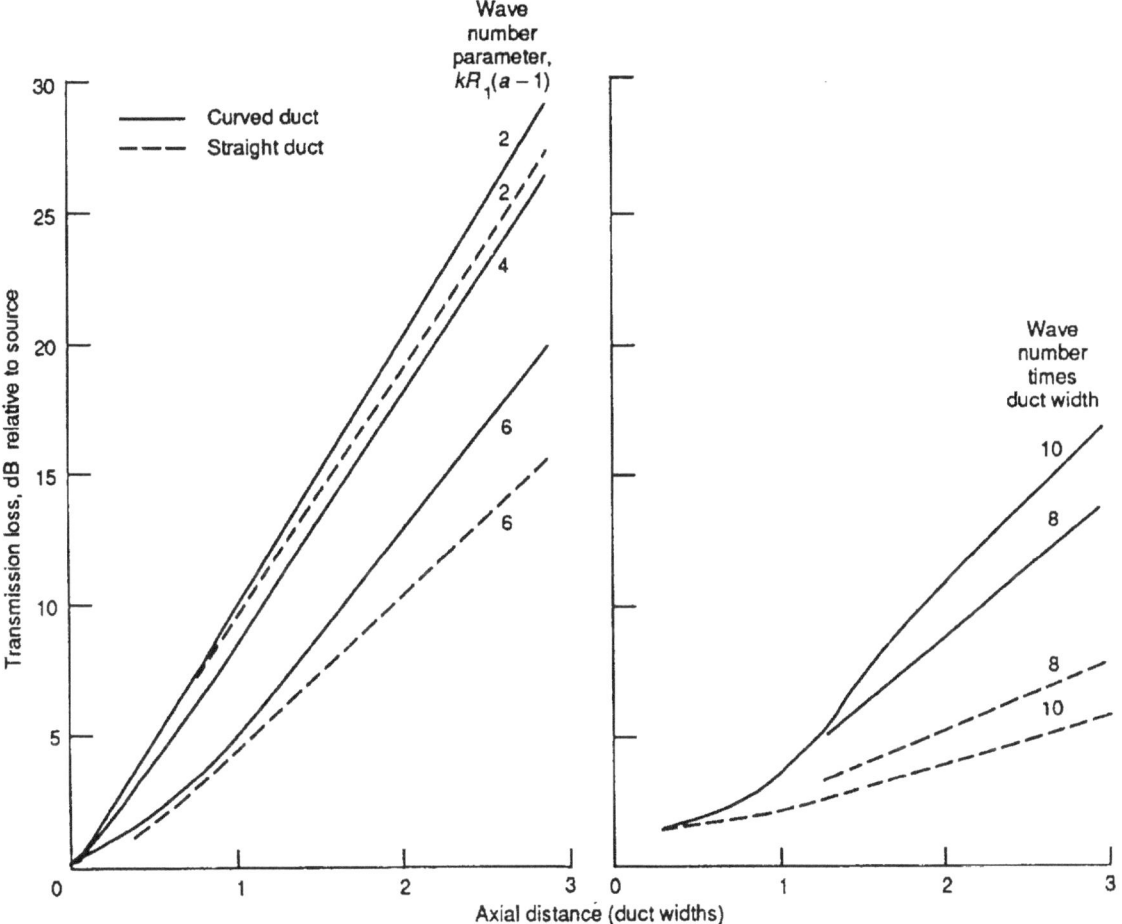

Figure 11.2.—Transmission loss in curved and straight ducts. Radius ratio, $a = 2$; nondimensional wall admittance, $\eta = 1 + i\,0.25$; From Myers and Mungur (1976).

First, they obtained extensive tables of complex eigenvalues for a selected range of parameters and for four modes of motion. With the help of several figures (figs. 11.3 to 11.9) they illustrate and present their results. For all the results (unless noted otherwise) the lining has nondimensional depth ($d_* = d/2R_2$) of 0.05, its specific acoustic resistance ($R_* = R/\rho c$) is 1.5, and generally the calculations pertain to 45° bends. Nondimensional frequency f_* equals $2f_o R_2/c$, or 22.32. Figure 11.3(a) shows the effect of radius ratio $a = R_2/R_1$ on sound attenuation. The sound attenuation of this fundamental mode (0,0) increases as the radius ratio decreases. This may be due to the increase in the wave path length for a fixed bend angle. When the nondimensional median arc length is fixed by varying the bend angle and the inner or outer radius (fig. 11.3(b)), the effect of the radius ratio on the sound attenuation becomes insignificant. The sound attenuation also increases as the bend angle increases (fig. 11.4) but decreases with broadening bandwidth as the specific acoustic resistance ratio R_* increases (fig. 11.5). The nondimensional frequency of the peak attenuation increases as the lining thickness decreases. This phenomenon, shown in both figures 11.5 and 11.6, was also well demonstrated in studies made of sound waves propagating in acoustically lined straight ducts with both rectangular and circular cross sections. The sound attenuation in curved ducts lined with dissimilar acoustic linings of two different depths differed greatly (fig. 11.7). For a fixed bend angle the sound attenuation for an outer-wall-lined duct is much greater than that for an inner-wall-lined duct (fig. 11.8). This may be due to the difference between the arc lengths of the outer and inner walls. As shown in figure 11.9 the fundamental mode ($n = 0$) is capable of traveling through the duct at all frequencies, but the higher wave modes ($n = 1, 2$, and 3) are capable of traveling through the duct only above their respective cutoff frequencies. Ko and Ho's (1977) main comments are as follows:

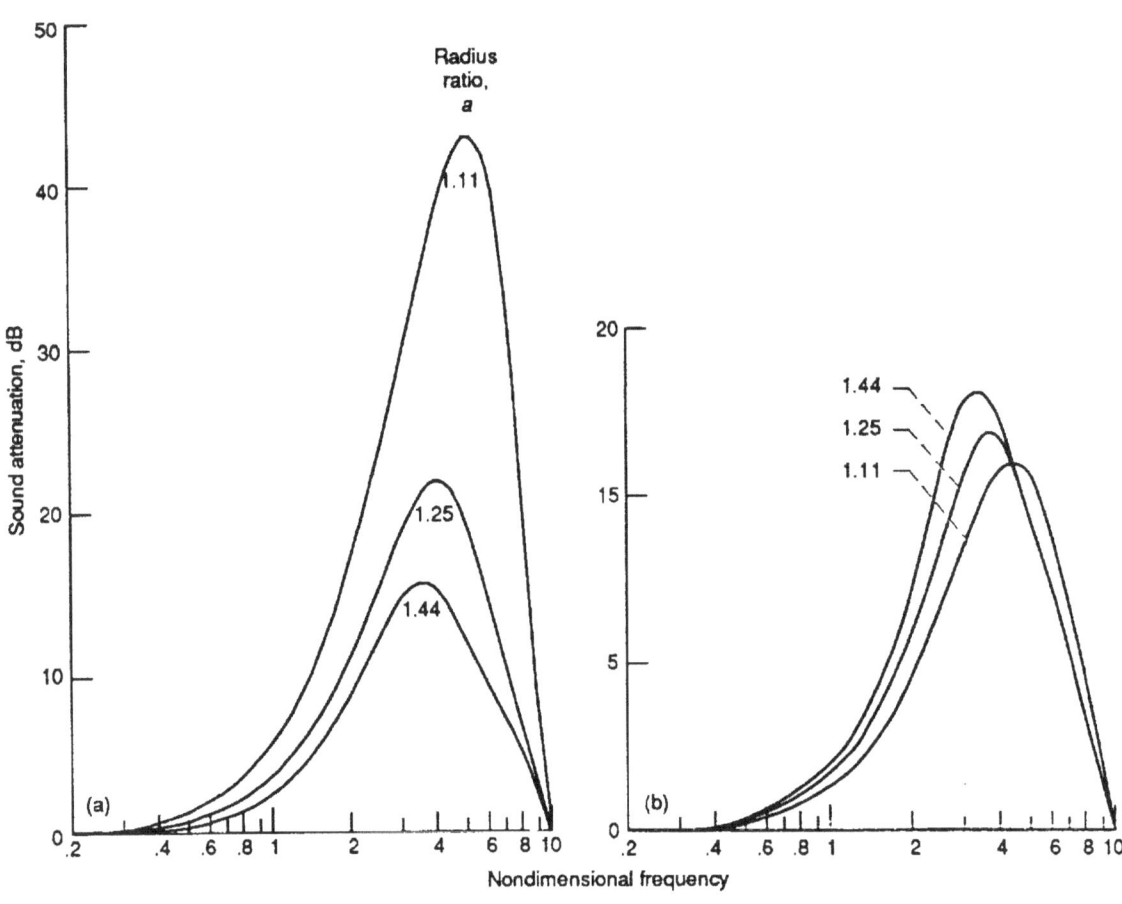

(a) Overall bend angle, $\Theta = 45°$.
(b) Nondimensional arc length, 2.

Figure 11.3.—Variation of sound attenuation with radius ratio. Specific acoustic resistance ratio, 1.5; nondimensional lining thickness on inner and outer walls, 0.05; nondimensional characteristic frequency, 22.32. From Ko and Ho (1977).

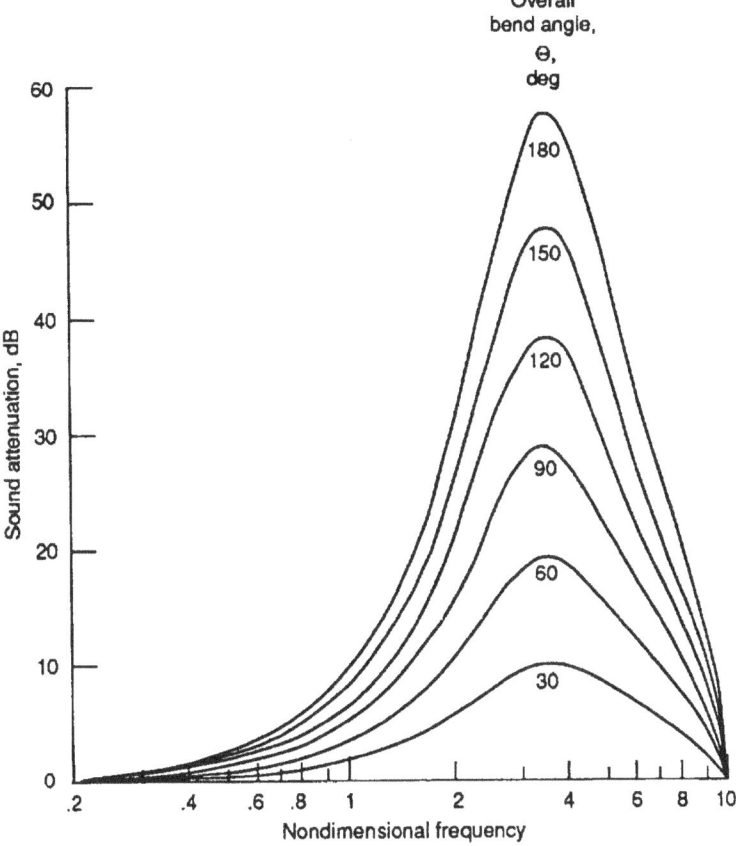

Figure 11.4.—Variation of sound attenuation with overall bend angle. Specific acoustic resistance ratio, 1.5; nondimensional lining thickness on inner and outer walls, 0.05; nondimensional characteristic frequency, 22.32; radius ratio, a = 1.43. From Ko and Ho (1977).

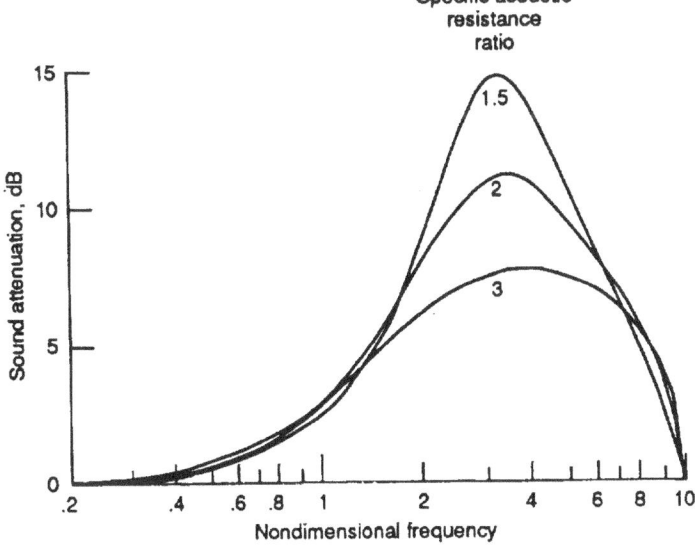

Figure 11.5.—Variation of sound attenuation with specific acoustic resistance ratio (same for inner and outer walls). Nondimensional lining thickness on inner and outer walls, 0.05; nondimensional characteristic frequency, 22.32; radius ratio, a = 1.43; overall bend angle, Θ = 45°. From Ko and Ho (1977).

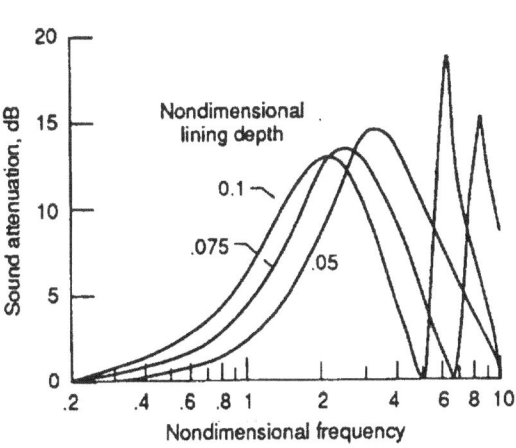

Figure 11.6.—Variation of sound attenuation with nondimensional lining depth. Specific acoustic resistance ratio, 1.5; nondimensional characteristic frequency, 22.32; radius ratio, a = 1.43; overall bend angle, Θ = 45°. From Ko and Ho (1977).

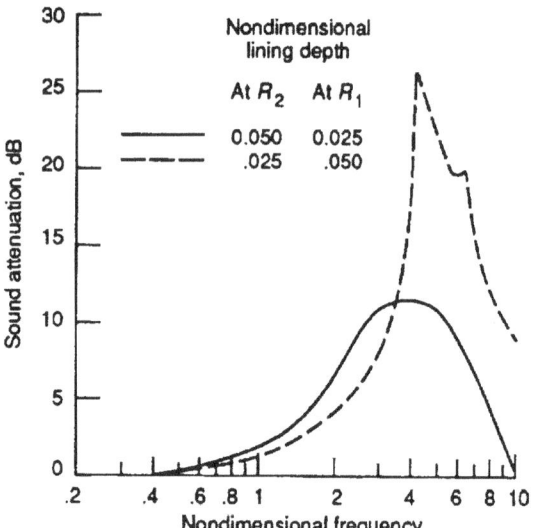

Figure 11.7.—Sound attenuation for two curved ducts with dissimilar acoustic linings. Specific acoustic resistance ratio, 1.5; nondimensional characteristic frequency, 22.32; radius ratio, a = 1.43; overall bend angle, Θ = 45°. From Ko and Ho (1977).

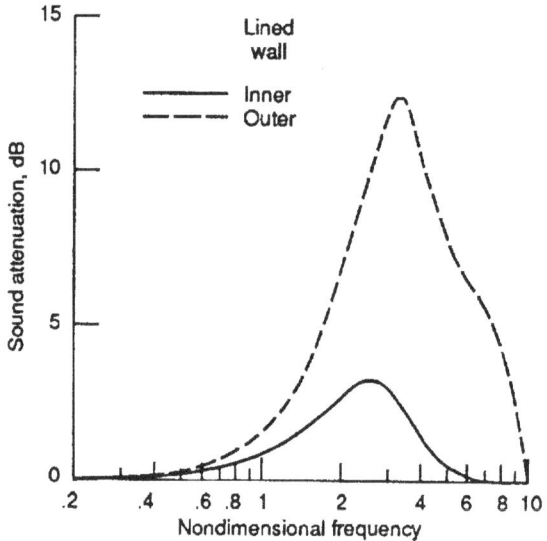

Figure 11.8.—Sound attenuation spectra for one-wall-lined ducts. Nondimensional lining depth, 0.05; nondimensional characteristic frequency, 22.32; radius ratio, a = 1.43; overall bend angle, Θ = 45°. From Ko and Ho (1977).

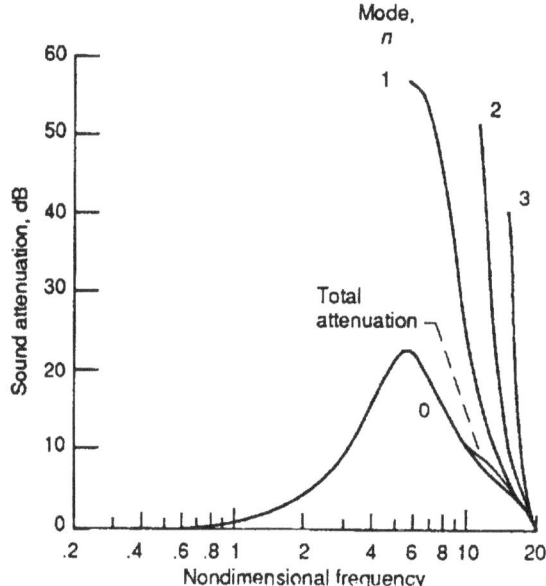

Figure 11.9.—Attenuation spectra for 0, 1, 2, and 3 modes and total attenuation. Specific acoustic resistance ratio, 1.5; nondimensional lining depth on inner and outer walls, 0.025 (0.5 in.); nondimensional characteristic frequency, 22.32 (characteristic frequency, 15 000 Hz; speed of sound, c = 1120 ft/s); radius ratio, a = 1.25 (R_1 = 8 in.; R_2 = 10 in.); overall bend angle, Θ = 45°. From Ko and Ho (1977).

The results obtained for a lined curved duct section, without terminations at either ends of the duct, are, in many respects, similar to those obtained for lined straight ducts. However, the characteristic of sound transmission in a lined duct system consisting of a bend joining straight duct sections may be different from that in a lined straight duct.

A different, extensive study on the propagation of sound waves in duct bends, acoustically lined on all four walls, has been done by Ko (1979). This is a rare three-dimensional study. The method is basically the same as that used by Ko and Ho (1977), but the equations are written without omitting the third dimension (the duct depth) and are extended by using as an additional variable the acoustical lining on the bend's roof and floor.

The eigenvalues of motion, the angular wave numbers of the propagation, were calculated by including in the computations wave numbers of motion in the vertical duct direction. The three-dimensional study shows that, generally, the effect of two or more acoustically lined walls is not negligible; the sound attenuation is increased. In one of his figures (fig. 11.10) Ko (1979) gives sound attenuation calculated for several different duct depths and a curve obtained by simpler two-dimensional calculations. He states that the limiting case of the three-dimensional case is the two-dimensional one (i.e., calculation of the sound attenuation based on the three-dimensional model approaches that based on the two-dimensional model, as it should).

Baumeister (1989) evaluates attenuation in an acoustically lined S-shaped duct. This is an unusual case because his lining is not typical; it is not the locally reactive type considered by all previous authors, but is rather an extended reaction liner that admits axial wave propagation in its material. His propagation theory and property formulas were validated by a number of experiments. Figure 11.11 shows the degradation of acoustic power flow in S-shaped ducts for one (representative) type of lining but for three duct lengths (i.e., for three degrees of duct offset). Increasing duct offset increases the attenuation of the transmitted power. Next, in figure 11.12, Baumeister explains the effect of lining thickness on power flow for the bend geometry shown in figure 11.11(b). It becomes obvious that only the lining layers immediately adjacent to the duct passage contribute to the attenuation of acoustic energy. Extremely thin layers of absorbing material do not attenuate waves to any significant degree. Finally, a useful map has been calculated (fig. 11.13) in which normalized acoustic attenuation contours for an S-shaped duct have been drawn. This is a design tool that can be obtained for any duct section.

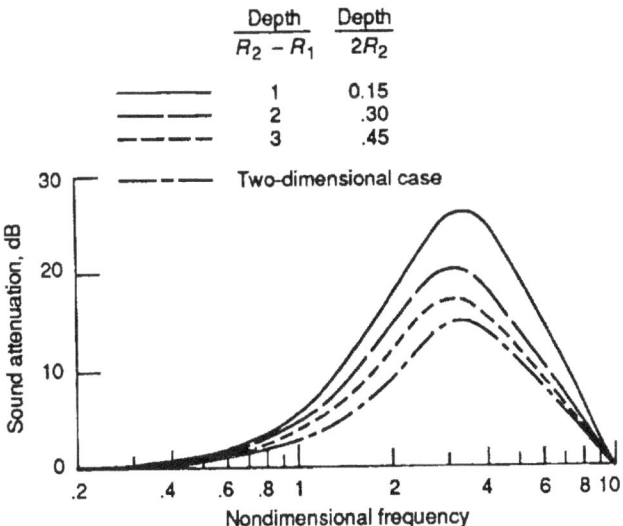

Figure 11.10.—Variation of sound attenuation with duct width. Specific acoustic resistance ratio, 1.5; nondimensional lining depth on inner and outer walls, 0.05; nondimensional characteristic frequency, 22.32; radius ratio, $a = 1.43$; overall bend angle, $\Theta = 45°$. From Ko (1979).

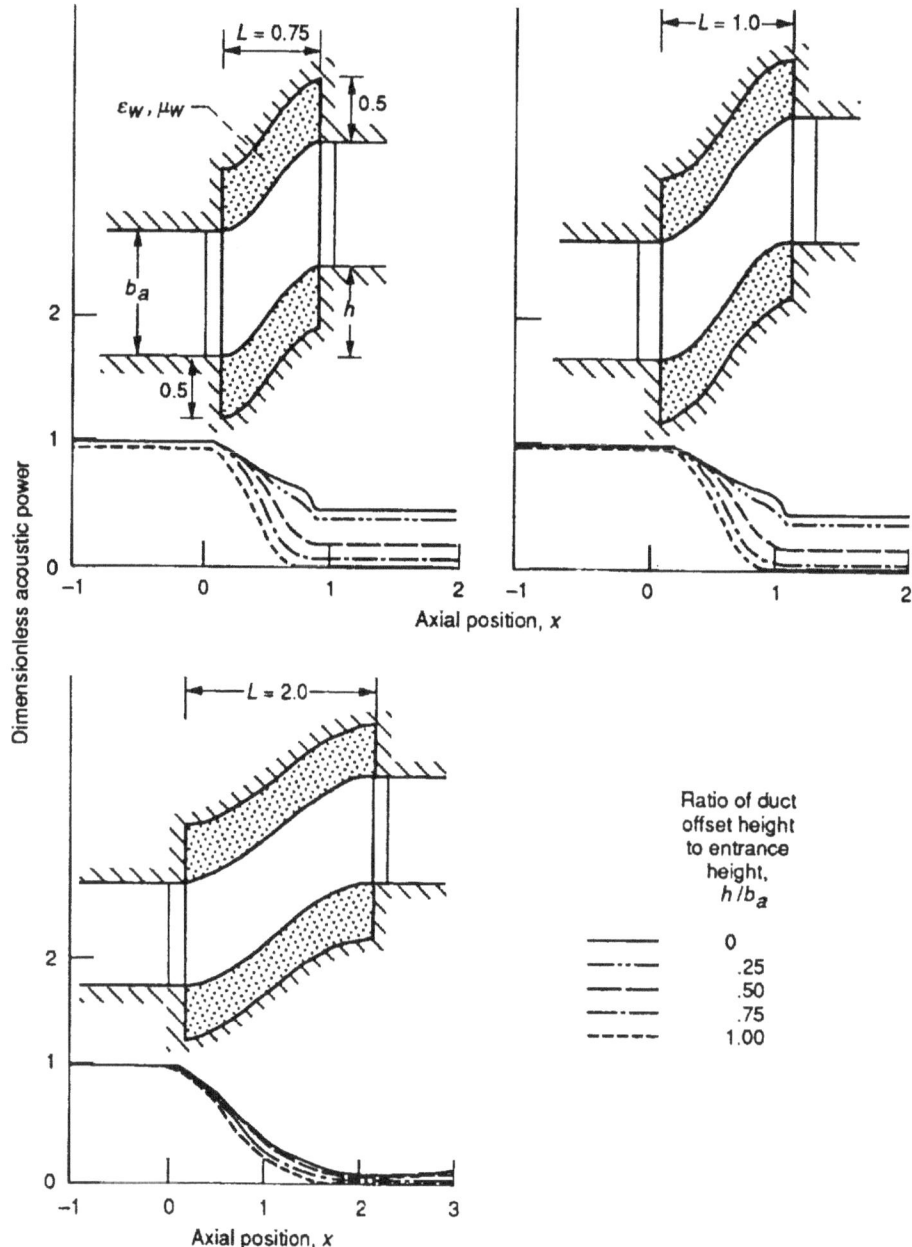

Figure 11.11.—Attenuation of transmitted power with duct offset for various dimensionless duct lengths L. Wall dimensionless complex property constants, $\epsilon_w = 1 - i\,2.83$ and $\mu_w = 4.1$; dimensionless frequency, $f = 1$. Baumeister (1989).

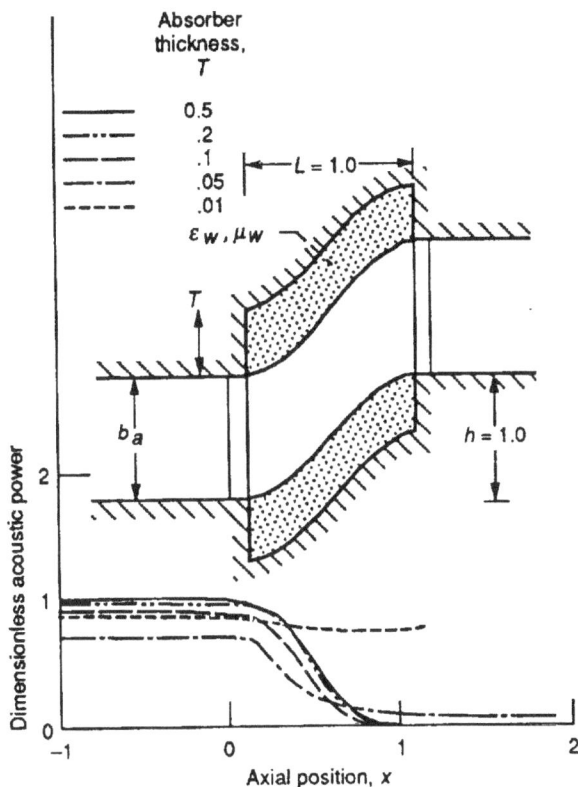

Figure 11.12.—Variation of magnitude of power in axial direction with absorber thickness. Wall dimensionless complex property constants, $\epsilon_w = 1 - i\ 2.83$ and $\mu_w = 4.1$; dimensionless frequency, $f = 1$. From Baumeister (1989).

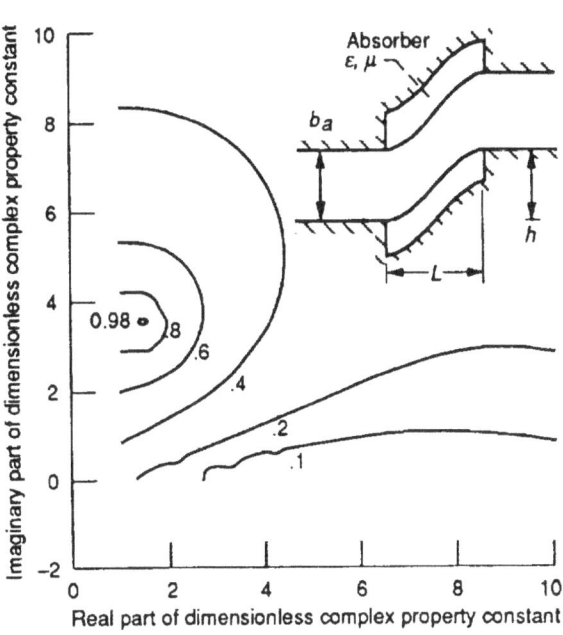

Figure 11.13.—Normalized acoustic attenuation contours for curved duct. Dimensionless frequency, $f = 1$; dimensionless duct length, $L = 1$; dimensionless duct offset height, $h = 1$; maximum attenuation, $dB_{max} = 36.5$; dimensionless complex property constants, $\epsilon_{opt} = 1.5 - i\ 3.5$ and $\mu = 4.1$. From Baumeister (1989).

12.0 Energy Flow

Cummings (1974) did not calculate energy flow in the bends that he tested, but he nevertheless outlines the procedure for such calculations. In an opening sentence he notes that in a hard-wall curved duct section without flow, the acoustic modes are orthogonal and thus the total energy flow is simply the sum of the contributions from various modes. (It is not necessary to recall here the standard equations for acoustic intensity and power in the angular direction.)

Two graphs (fig. 12.1) by Myers and Mungur (1976) merit attention. Figure 12.1(a) shows, for a hard-wall bend with $a = 2$, the distribution of angular acoustic intensity I_θ for several angular stations in the bend, including at the inlet ($\theta = 0$). Intensity peaks occur near the outer wall. Figure 12.1(b) shows the same distribution for the same duct but with an acoustical lining. The axial sound intensity attains its maximum value near the center of the duct. The absorbing walls smooth out the energy distribution between the radial walls.

Rostafinski (1974b) calculated sound intensity in a range of frequencies for hard-wall duct bends of three radius ratios. The values he obtained for energy flux are compared in figure 12.2 with energy flux in straight ducts, which has the well-known form $pv = \rho c v_c^2/2$. The figure shows that the ability of bends to transmit acoustic energy depends strongly on the frequency and the bend radius ratio. Rapid decreases in the transmissivity of bends sharper than $a = 1.5$ should be linked to the cutoff characteristics evaluated in section 9.0.

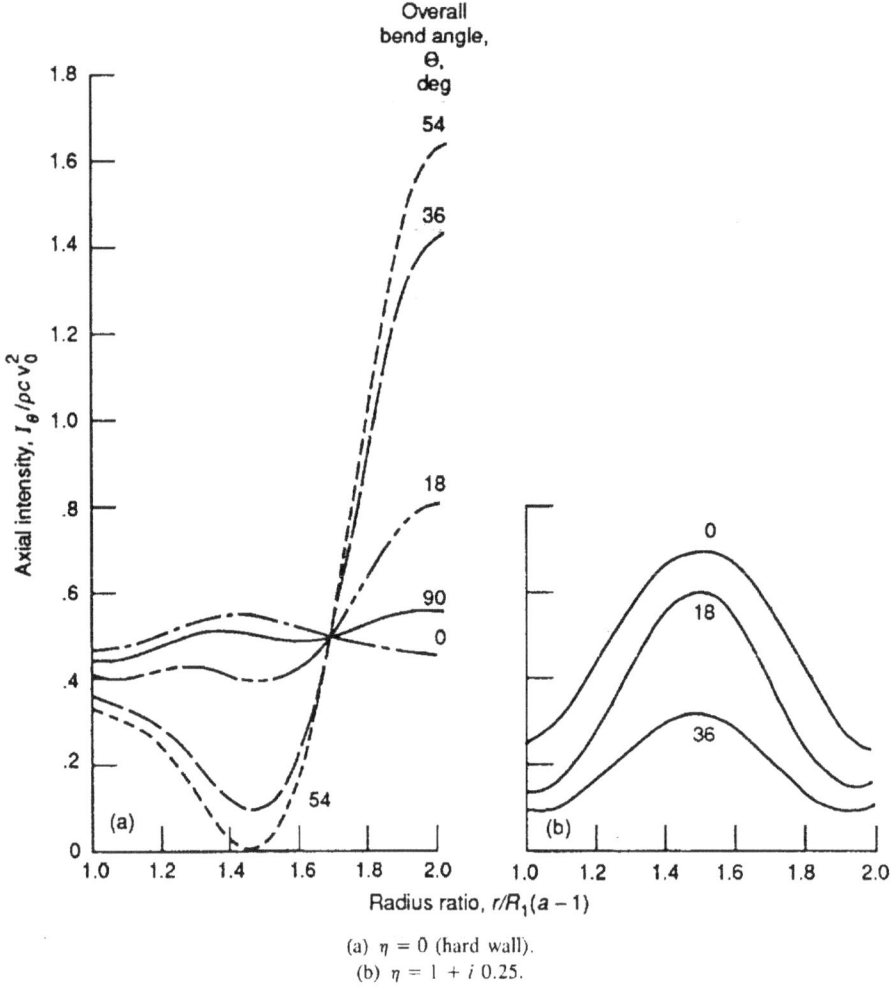

(a) $\eta = 0$ (hard wall).
(b) $\eta = 1 + i\,0.25$.

Figure 12.1.—Axial intensity distributions for two nondimensional wall admittances η. Radius ratio, $a = 2$; wave number parameter, $kR_1(a - 1) = 6$. From Myers and Mungur (1976).

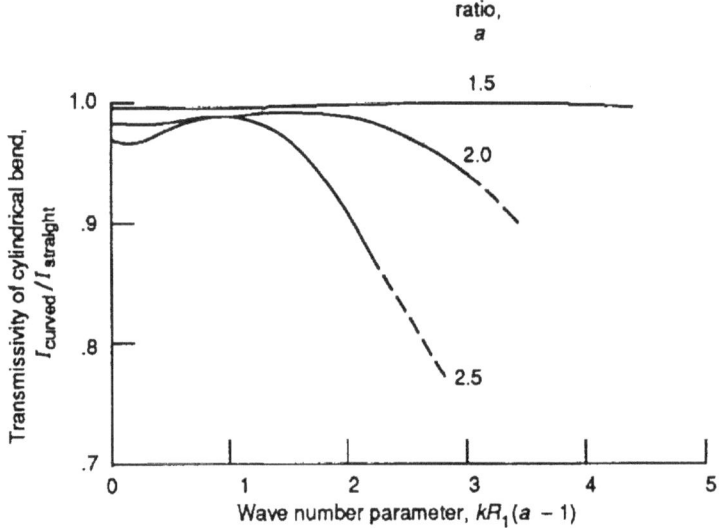

Figure 12.2.—Transmission of wave energy in curved ducts. From Rostafinski (1974b).

13.0 Bends With Turning Vanes

An axial partition in a cylindrical bend divides it into two concentric bends and thus creates a completely new boundary condition for the propagating sound waves. Such a partition significantly alters the sound transmission through the bend. Three papers deal with this problem, and in all three there is excellent agreement between analytical and experimental results.

Fuller and Bies (1978a) analytically and experimentally evaluated an acoustical discontinuity created by inserting a partition along the centerline of a 180° bend. The partition drastically altered the acoustic properties of the bend because it divided it into two bends of unequal path length. Fuller and Bies limited their analysis to frequencies lower than the cutoff frequency of the (1,0) mode in two straight ducts connected by a bend. Their analysis began with formulation of the characteristic equation (separation of variables) followed by a set of equations for each discontinuity (i.e., at the inlet and outlet of the 180° bend). Derivation of Fourier coefficients for each mode allowed the propagating mode and an infinite set of evanescent waves to be evaluated. Numerical application of the derived equations and experiments was done for a bend of $a = 2$ divided, at the centerline, into two concentric bends by a thin, rigid partition. The description of test procedures and results along with their interpretations are given here as written by Fuller and Bies (1978a):

> Three parameters were measured. The power reflection coefficient, an indication of how much sound is reflected back towards the source, and the characteristic impedance, an indication of how severe a discontinuity the bend presents to acoustic propagation, were obtained by measuring the standing wave in the upstream straight duct and applying standing-wave theory. Values of the experimental reflection coefficient plotted against a nondimensional frequency parameter $[kR_1(a - 1)]$ are shown in Fig. [13.1]. Since the analysis is limited to less than the cut-off frequency of the (1,0) mode in the straight-duct section, for which $[kR_1(a - 1)] = \pi$, values of $[kR_1(a - 1)]$ are terminated at $[kR_1(a - 1)] = 3.02$. Experimental values of the resistive impedance $R/\rho c$ are shown in Fig. [13.2] while those of the reactive part $X/\rho c$ are shown in Fig. [13.3].
>
> The insertion loss [is] a measure of the attenuation in decibels of the incident wave.... The insertion loss is defined to be the difference in decibels of $P_{00}^i - E_{00}^i$. Thus using standing-wave theory it can be shown that,

$$P_{00}^{\max} = \left(1 + \alpha_r^{1/2}\right) P_{00}^i,$$

$$\therefore P_{00}^{\max}(\text{dB}) = 20 \log_{10}\left(1 + \alpha_r^{1/2}\right) P_{00}^i(\text{dB}),$$

$$\therefore P_{00}^i = P_{00}^{\max} - 20 \log_{10}\left(1 + \alpha_r^{1/2}\right).$$

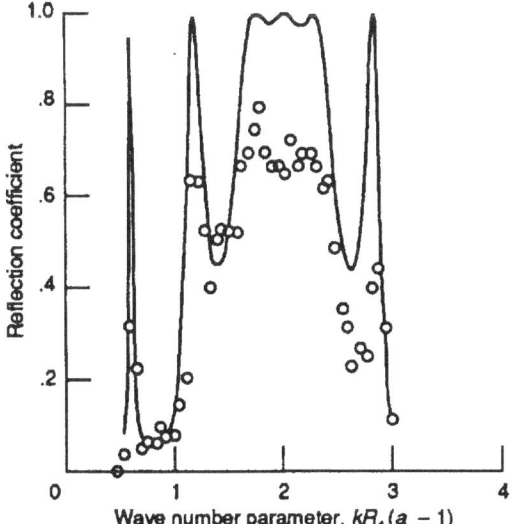

Figure 13.1.—Sound power reflection coefficient of compound bend. From Fuller and Bies (1978a).

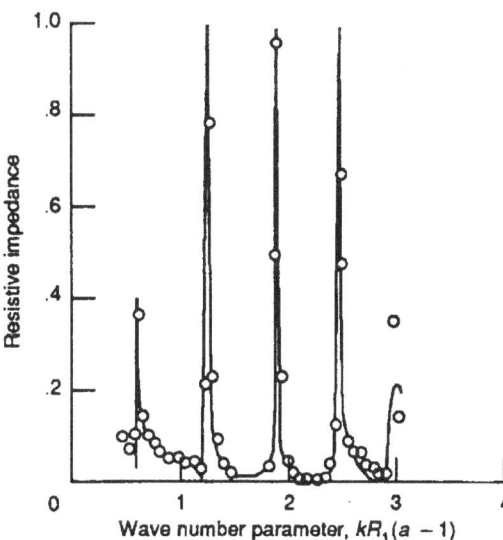

Figure 13.2.—Resistive part of characteristic impedance of compound bend. From Fuller and Bies (1978a).

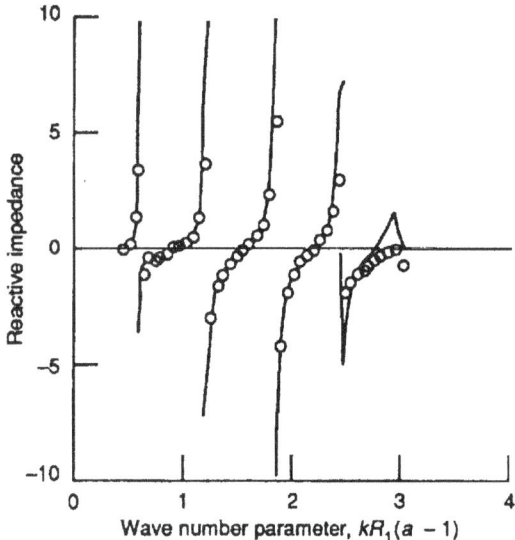

Figure 13.3.—Reactive part of characteristic impedance of compound bend. From Fuller and Bies (1978a).

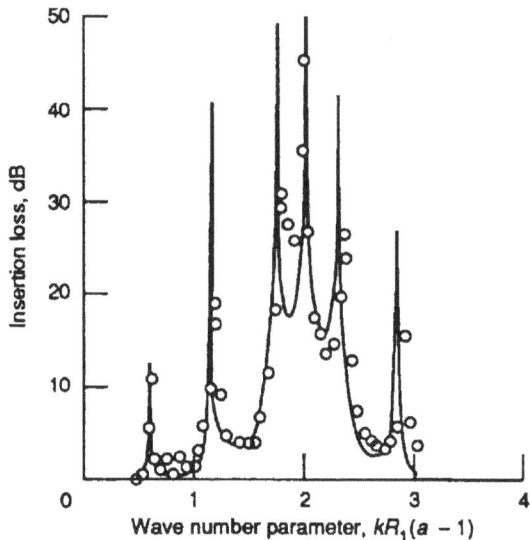

Figure 13.4.—Insertion loss of compound bend. From Fuller and Bies (1978a).

Hence the insertion loss of the bend is

$$\text{measured insertion loss (dB)} = P_{00}^{\max} - 20\,\log_{10}\left(1 + \alpha_r^{1/2}\right) - E_{00}^i (\text{dB})$$

[where] α_r is the measured power reflection coefficient, [superscript i refers to the incident wave, E is the measured sound pressure level in the downstream duct, and $P^{\max}$ is the measured maximum sound pressure in the upstream duct.]

The theoretical insertion loss is

$$\text{I.L.} = -20\,\log_{10}(1 - \alpha_r)^{1/2}\ldots.$$

As can be seen in Fig. [13.1] there is close agreement between theoretical and experimentally measured values of the frequencies at which maxima occur....

The magnitude of the experimental reflection coefficient is consistently less than predicted at the maxima. This was thought to be due to the difficulty of determining accurately the standing wave ratio when it is quite large....

As can be seen in Fig. [13.1] a curved 180° bend with a partition positioned on its centreline provides a large disruption to sound propagation. In fact the theory developed here predicts that at a number of the dimensionless frequencies...[$kR_1(a - 1)$]...the power reflection coefficient is very close to unity....

Theoretical and experimental values of the resistive and reactive parts of the characteristic impedance are in good agreement as shown in Figs. [13.2] and [13.3]...

Much closer agreement is demonstrated by comparison of experimental and theoretical values of insertion loss shown in Fig. [13.4] than reflection coefficient shown previously in Fig. [13.1]. This observation supports the argument presented previously to explain the discrepancies shown in the latter figure....

The partition was found to significantly alter the sound propagation through the bend, resulting in high reflection of sound at a number of discrete frequencies.

Cabelli (1980) extensively studied the effects of a partition in a bend. He checked his analytical findings by experiments on 45°, 90°, 150°, and 180° bends with axial vanes of different lengths extending through the bend but, generally, not reaching the junction sections between the straight ducts and the ends of each bend. The vane length dictated to a large degree the extent of the reflections that occurred.

Citing Cabelli, figures 13.5 and 13.6 show the behavior of the reflection coefficient for five bends and a variety of turning vanes. For the range of geometries considered in this work, the difference in length between the two sound paths in the curved section could be equal to half the wavelength only for a bend

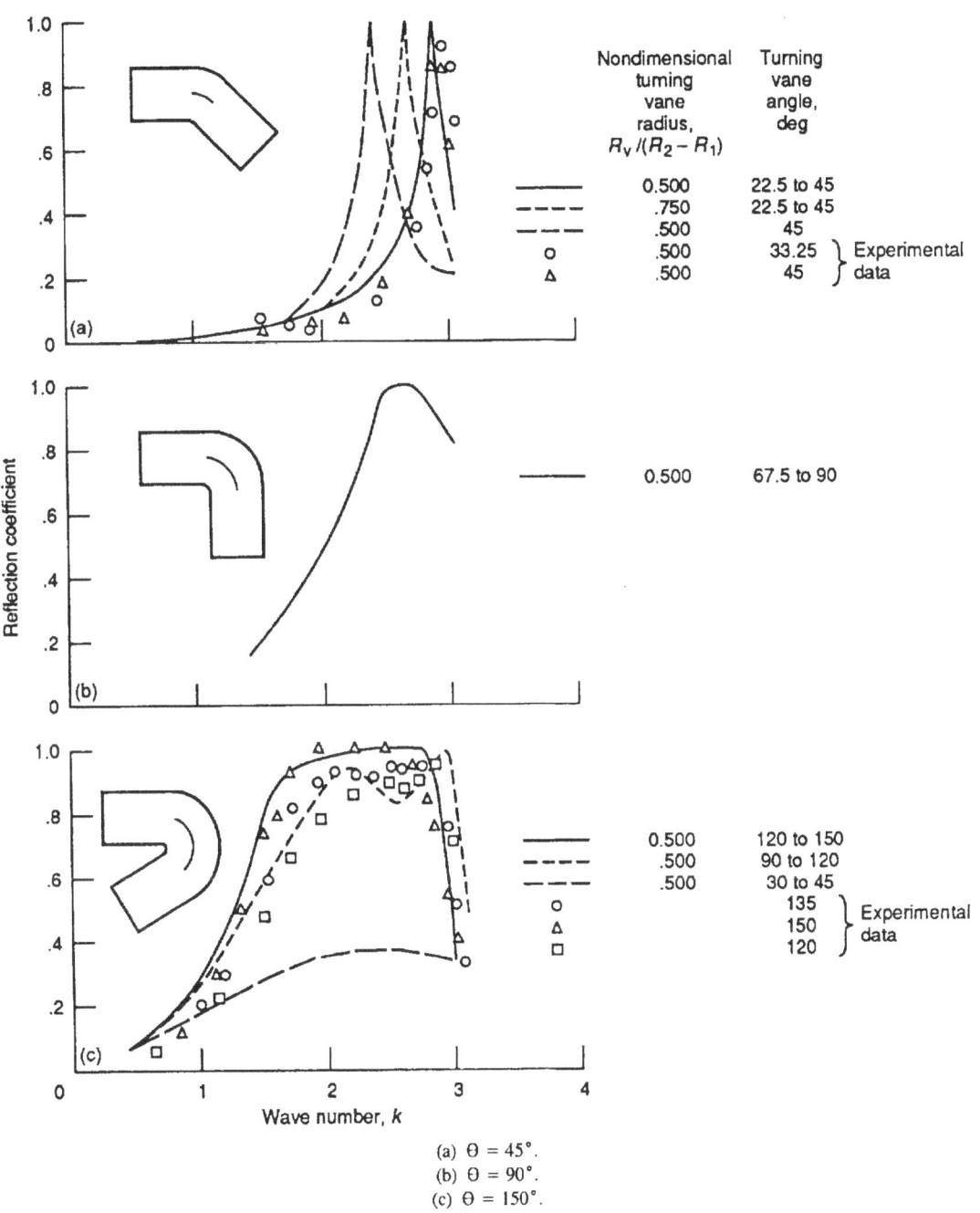

(a) Θ = 45°.
(b) Θ = 90°.
(c) Θ = 150°.

Figure 13.5.—Variation of reflection coefficient with wave number for three overall bend angles Θ, radius ratio $a = 9$, and turning vane located in bend. From Cabelli (1980).

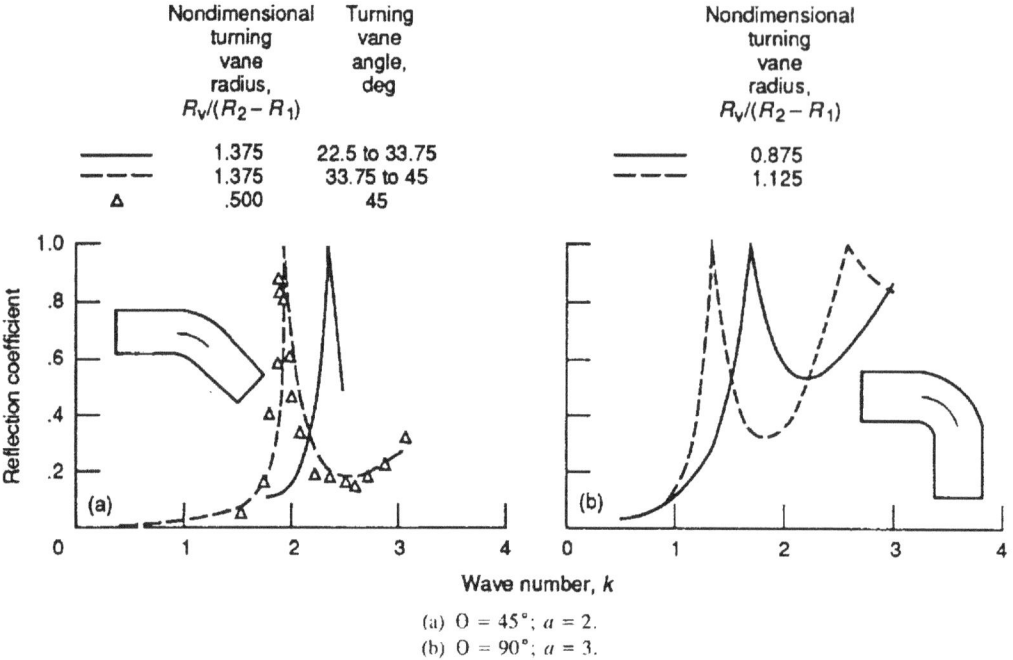

Nondimensional turning vane radius, $R_v/(R_2 - R_1)$	Turning vane angle, deg	Nondimensional turning vane radius, $R_v/(R_2 - R_1)$
——— 1.375	22.5 to 33.75	——— 0.875
— — — 1.375	33.75 to 45	— — — 1.125
△ .500	45	

(a) $O = 45°$; $a = 2$.
(b) $O = 90°$; $a = 3$.

Figure 13.6.—Variation of reflection coefficient with wave number when turning vane is located in bend for two bend radius ratios. From Cabelli (1980).

angle of 150°. With the turning vane angle also equal to 150° the mean path length difference was 1.31, which corresponds to total reflection for a wave number of 2.4. With a turning vane of 120° the path length difference for total cancellation corresponds to a wave number of 3.0. The experimental and numerical characteristics given in figure 13.5 predict and verify these trends. As Cabelli (1980) states,

> The results of Figure [13.5] also describe a shift of the high reflection zone toward lower frequencies as the length of the turning vane was increased. When the angle of the bend was equal to 45°, this was achieved either by locating the turning vane further out in the bend (i.e., $R_v = 0.750$) [in other words, increasing the radius of the vane R_v] or literally by increasing the length of the vane at a given radius. In the former case, the mean path length difference remained unchanged although the individual mean path lengths were longer.... Another feature of the results which can be seen in Figure [13.5] is the broadening of the high reflection band as the angle of the bend was increased. This however, was found to be a function of the tightness of the bend. For bends of smaller curvature and with identical path length differences, the reflection coefficient retained the narrow selective characteristic seen in Figure [13.5] but developed additional peaks of low transmission at higher angles. Figure [13.6], which describes the reflection coefficient for a 45° bend with an inner radius of unity and for a 90° bend with an inner radius of 0.5, provides an example of this behaviour. It appears that multiple peaks are characteristic of bends with small curvatures whereas tight bends such as the 150° bend of Figure [13.5] will generate broad reflection patterns. This is confirmed by solutions for a 180° bend with a 2:1 radius ratio which are discussed later and which also produced peaks of reflection at discrete frequencies.
>
> The presence of turning vanes in duct bends generally gave rise to cross modes which were of much greater amplitudes than those found in the absence of centre bodies....
>
> The numerical solution method was also used to simulate the experimental conditions of Fuller and Bies [1978a]. The input data described a 180° bend with a radius of 2:1.... Turning vanes were positioned either at $R_v = 1.375$ or at $R_v = 1.65$ and their end co-ordinates were either (8,26) corresponding to an angle $160° \leq \alpha < 180°$ or (8,27) describing an angle of 180° with straight sections at each end. These limits define a range which covers the geometry investigated analytically and experimentally by Fuller and Bies [1978a]. The numerical results for the reflection coefficient were generally in good agreement with the experimental results. Both the numerical solution and the modal analysis of Fuller and Bies [1978a] predicted sharp peaks in the reflection coefficient at values of the wave number parameter approximately equal to 0.7, 1.2 and 3. These were substantiated by the trends observed experimentally. The length and location of the turning vane

were found to have a small influence on the behaviour described by the numerical solution, the most evident effect being a change in the position of the peaks. However, one aspect of the numerical solution which was not observed in the modal solution is concerned with the behaviour at values of the wave number between 1.8 and 2.2 approximately. In this range, Fuller and Bies predicted nearly total cancellation of the transmitted sound whilst the experimental results and the present numerical solution indicated values of the (pressure) reflection coefficient nearer 0.8. This discrepancy was probably caused by the presence of higher order cross modes at the discontinuities. The influence of cross modes higher than the first was neglected in the matching boundary condition used for the analytical solution [given by Fuller and Bies (1978a)]. In fact, the numerical solution predicted amplitudes of these evanescent modes which were higher when the wave number was equal to 2.0 than for corresponding solutions with wave number parameters equal to 1.0 or 2.75. A typical solution indicated the normalized magnitude of the second cross mode to be equal to 0.40 when k was equal to 2.0. When the wave number parameter was equal to 1.0 and 2.75, the normalized magnitude of the second order mode was equal to 0.07 and 0.12 respectively. Figure [13.7] displays some results obtained by the numerical method of solution superimposed on the theoretical and experimental results of Fuller and Bies [1978a]. The power reflection coefficient plotted in the figure is equal to the square of the pressure reflection coefficient R_c. The numerical results for $R_v = 1.375$ and $R_v = 1.65$ can be seen generally to straddle the experimental results of [Fuller and Bies] (which were obtained for a value $R_v = 1.50$) and the deviation from the experimental results is smaller than that obtained by Fuller and Bies for the range of values of the wave number in which the second cross modes have significant magnitudes at the geometric discontinuities.

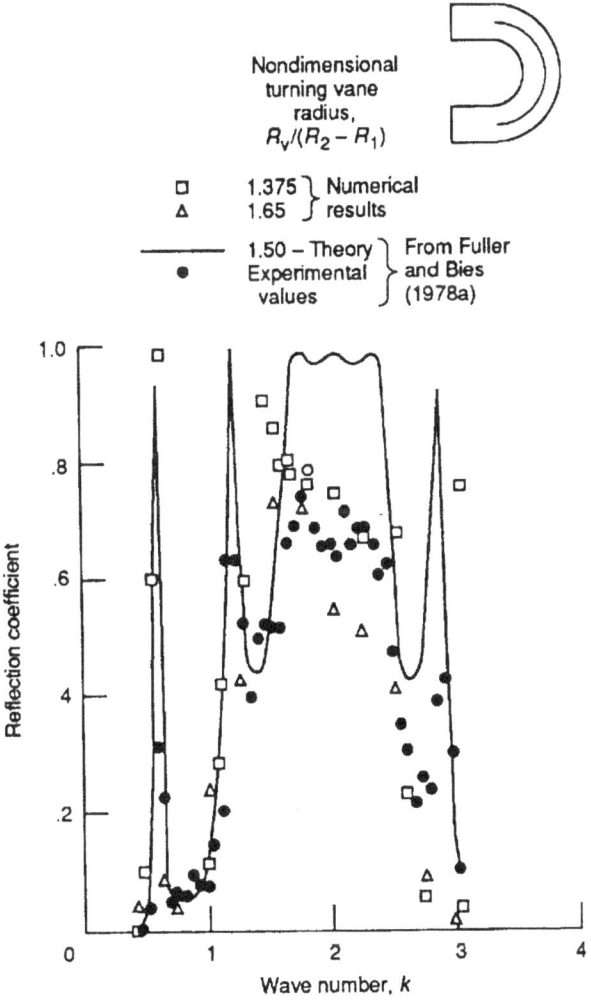

Figure 13.7.—Sound power reflection coefficient for 180° bend. Radius ratio, $a = 2$. From Cabelli (1980).

The first (documented) attempt to evaluate the effectiveness (in attenuating noise propagating in ventilation ducting) of a shaped partition in a duct bend was published by Luxton (1968). He showed, by experiments, that a properly designed partition that divides a bend into two paths differing in length by a half-wavelength does attenuate sound as expected.

A similar type of partition was designed and evaluated by Fuller and Bies (1978b). It is a crescent-shaped centerbody (shown in fig. 13.8) set in a bend connecting two straight ducts. The authors call this device an attenuator because it attenuates sound effectively in the design frequency range. As in other studies the analytical and experimental procedure limited the propagating modes in the bend to the fundamental (0,0) mode. For convenience the authors adopted a reference amplitude of the incident wave as $P^i_{00} = 1 - 0i$; the design frequency for the propagating mode was 844 Hz. As in the study on the effects on sound propagation of a simple partition placed in a bend (Fuller and Bies, 1978b) this analysis also involved evaluating the sound power transmission coefficient and the transmission loss.

...The predicted values of transmission coefficient were then evaluated from

$$\alpha_t = 1 - \left| P^r_{00}/P^i_{00} \right|^2.$$

The theoretical values obtained for the original attenuator are shown in Figure [13.9(a)], where sound power transmission coefficient is plotted against a non-dimensional frequency parameter $[kR_1]$. Measured values are also shown in the figure for comparison.

As shown in Figure [13.9(a)] close agreement is observed between the predicted and measured frequencies at which minima in the transmission coefficient are observed, especially at lower frequencies. Slight discrepancies between the theoretical and experimental frequencies of minimum transmission are thought to be due to dimensional inaccuracies in the geometry of the attenuator, affecting the mean path difference between the two ducts. At low frequencies the wavelength of the incident sound is very much larger than the duct's small scale dimensions and thus only a small discrepancy results. However, at high frequencies this is no longer true and the discrepancies are larger.

The magnitude of the measured transmission coefficient, which agrees closely with that predicted at low frequencies, is progressively greater than predicted at increasing frequencies. Three possible reasons for this observation are suggested, as follows.

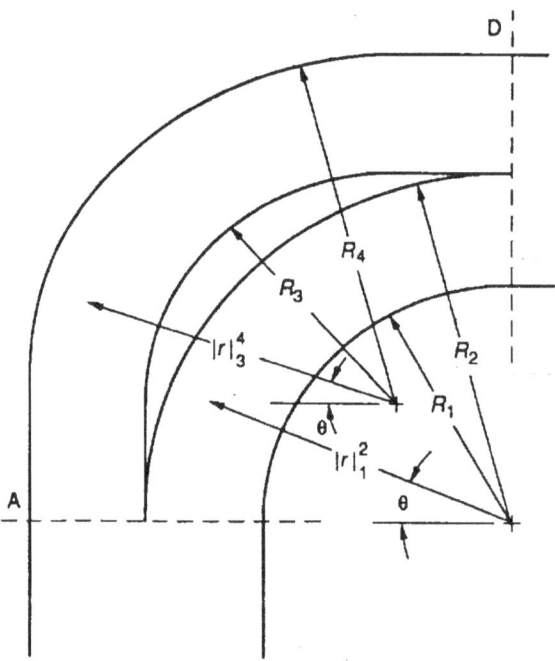

Figure 13.8.—Arrangement and coordinate system of attenuator. From Fuller and Bies (1978b).

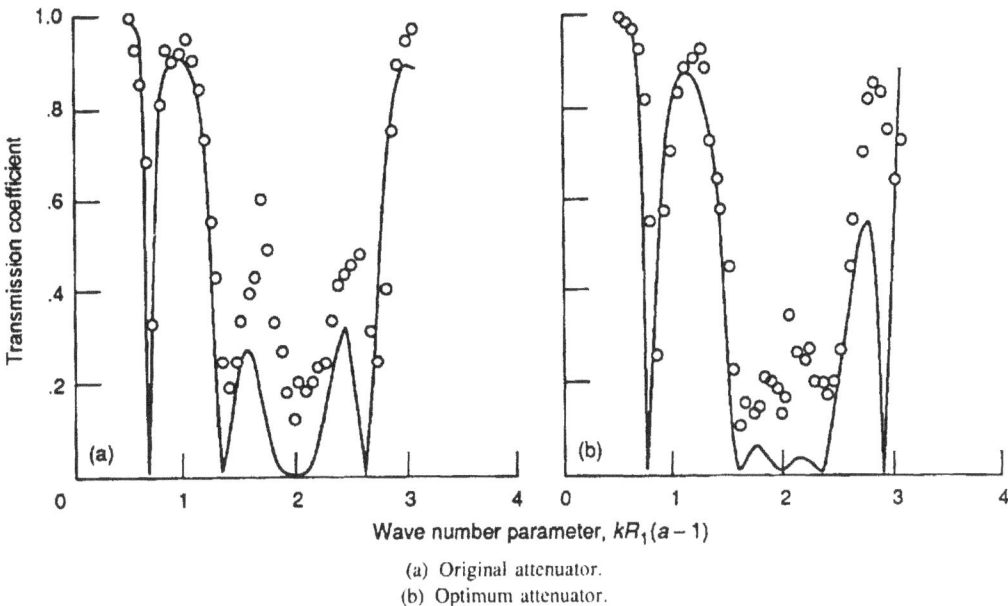

(a) Original attenuator.
(b) Optimum attenuator.

Figure 13.9.—Power transmission coefficients. From Fuller and Bies (1978b).

(1) The walls of the experimental duct are not absolutely rigid as supposed by theory. In fact they were found to vibrate and radiate sound.... (2) The minima of the standing wave measured in the upstream duct become sharper with increasing frequency, particularly at large values of the standing wave ratio n. This leads to error in evaluating the magnitude of the minimum pressure and results in a higher value of transmission coefficient than predicted. (3) The duct dimensions are not exactly described by the theory.

Minimum transmission at the design frequency may be fully accounted for in terms of reflection at the bend exit plane, interface D of Figure [13.8].... The additional minima are due to multi-reflections at interfaces A and D and are fully accounted for by the more exact theory presented here. The frequencies at which additional minima occur depend upon the magnitude of the path difference relative to the mean lengths of either of the ducts in the compound bend. For convenience we will take the inside duct mean length as the standard length for comparison. Thus the ratio of the inside duct length to path difference determines the frequencies of additional minima....

An attenuator designed for optimal attenuation characteristics has the following dimensions: with reference to Figure [13.8] its radii are $R_1 = 0.184$ meters, $R_2 = 0.248$ meters, $R_3 = 0.006$ meters and $R_4 = 0.070$ meters.

The theoretical and experimentally measured transmission coefficient values of this attenuator are shown in Figure [13.9(b)]. It can be seen that the extra minima have indeed moved closer to the design frequency than those shown in Figure [13.9(a)]. In the model attenuator a rejection band of 430 Hz centered at a design frequency of 844 Hz has been achieved.

Closer agreement is obtained in the position of theoretical and experimentally measured minima produced by the optimum attenuator. This is due to more accurate machining of components in the attenuator, thus achieving the correct mean path lengths.

The theoretical transmission loss is predicted by

$$\text{transmission loss} = -10 \log(\alpha_t).$$

Theoretical and experimentally measured values of transmission loss for the original attenuator are shown in Figure [13.10(a)] while those for the optimum attenuator are shown in Figure [13.10(b)]....

The attenuator provides large transmission loss at a series of discrete frequencies all below the cut-off frequency for the first cross-mode in the duct. The presence of these discrete frequencies is explained by reflection of incident sound at the exit and entrance to the attenuator. Their relative distribution is determined by the geometry of the attenuator.

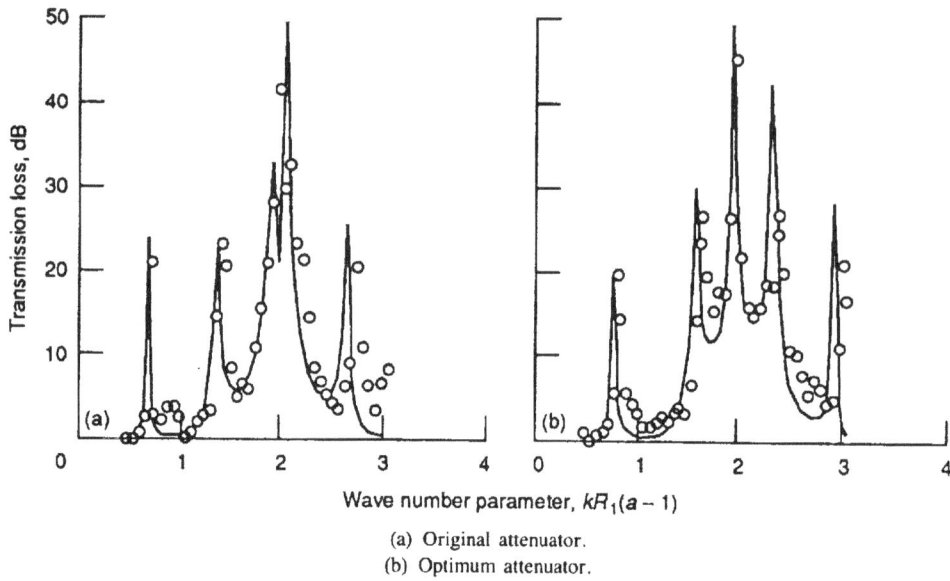

(a) Original attenuator.
(b) Optimum attenuator.

Figure 13.10.—Transmission loss. From Fuller and Bies (1978b).

14.0 Sound Propagation in Pipe Bends

Cummings (1974) supplements his study on the transmissivity of cylindrical bends in rectangular ducts by considerations on the acoustical characteristics of pipe bends. The problem is, he indicates, that the toroidal coordinate system required to suit the boundary conditions of a circular-section curved duct (pipe) is not one of the 11 coordinate systems in which the wave equation is separable. By using some numerical methods, however, it should be possible to evaluate the sound fields in pipe bends and calculate their characteristics. As a guide for future research in this area, some experimental data published by Cummings (1974) are given here. His descriptions and comments are as follows:

> ...In this paper, measurements are given of the sound field in a circular section curved duct, and certain plausible proposals are made regarding a design guide. These are supported by measurements. Figure [14.1] shows the measured sound pressure field in a circular section curved duct with pipe radius [r_2] = 0.0445 m, centreline bend radius = 0.445 m [k_0 = 35.1; and Θ = 90°]. It is seen that

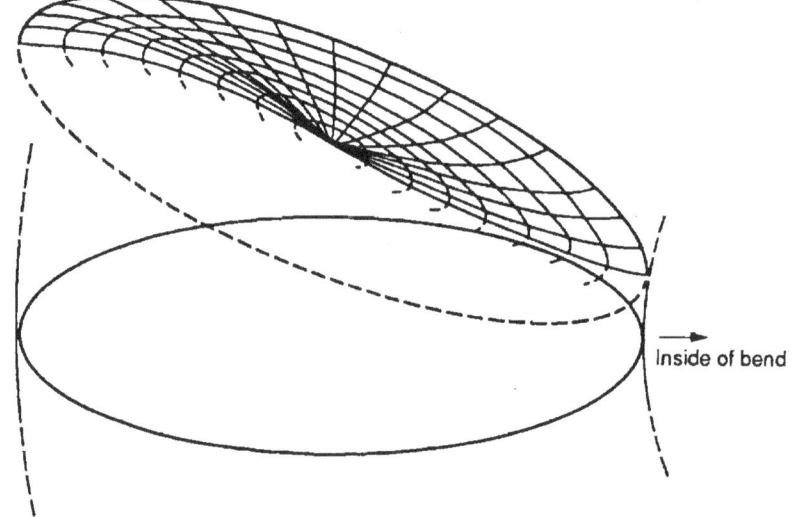

Inside of bend

Figure 14.1.—Sound field in circular-section curved duct. Imposed wave number, k = 35.1; centerline bend radius, R_m = 0.445 m; outer radius of pipe cross section, r_2 = 0.0445 m; Θ = 90°. From Cummings (1974).

the sound field appears very much the same as a circular portion of the sound field in a square section curved duct if the side of the square section is equal to the diameter of the circular duct, and the centreline radius is the same in both cases. That is, the sound pressure variation is principally in a direction parallel to a diameter of the bend; there is very little sound pressure variation in a direction at right angles to this in the plane of the duct cross-section.

Figure [14.2] shows a comparison of the sound pressure patterns, in the two directions previously mentioned, between the measurements in the circular section duct and predictions for a square section curved duct of width and depth equal to the pipe diameter, and the same centreline radius. The sound pressure pattern along the line AB is almost identical in the two cases (the discrepancy is within experimental error). Along the line CD, the sound pressure in the square duct is, of course, uniform since k_z has been assumed zero, while there is a small variation in the sound pressure in the circular section duct....

Figure [14.3] shows impedance measurements and predictions on the circular duct depicted in Figure [14.2]. In the nomenclature of the diagram in Figure [14.3], $(\ell_1 + \ell_2)$ in this case was equal to 0.1135 m. Three theoretical curves are shown: one calculated on the basis of an equivalent duct width of $\pi/4$ times the duct diameter, one on the assumption of an equivalent duct width equal to the duct diameter, and one derived by using the centreline length of the curved section. The first of these appears to be in best agreement with the measurements. The last of these is in almost as good agreement as the first, while the second is in poor agreement.

It is evident that the equivalent duct width of $\pi/4$ times the diameter gives good agreement with experiment. This can be used as a basis of a design guide; the ensuing procedure is simply that for a rectangular section duct.

In view of the mathematical difficulties in handling toroidal coordinates, Keefe and Benade (1983)

...propose an approximate model in which the flow through the circular cross section is considered to be a superposition of flows through rectangular slices which are stacked on top of one another to build up the circle. Cummings [1974]...has suggested the possibility of such an approach. We imagine that the circular cross section is split up into a collection of rectangular slices, each lying in a plane perpendicular to the axis of curvature having a height $H = dz$. We assume that the flow in each slice is that calculated from the velocity potential...for a duct of this shape and size. In other words, we assume that the rectangular slices do not communicate with one another, so that only negligible effects are produced by pressure variations in the vertical z direction and by shear between moving elements at the interslice boundaries.

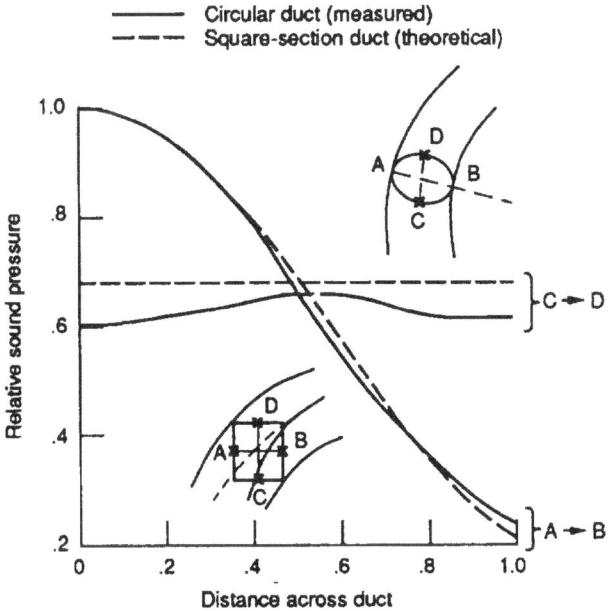

Figure 14.2.—Sound pressure distributions in circular and rectangular ducts. From Cummings (1974).

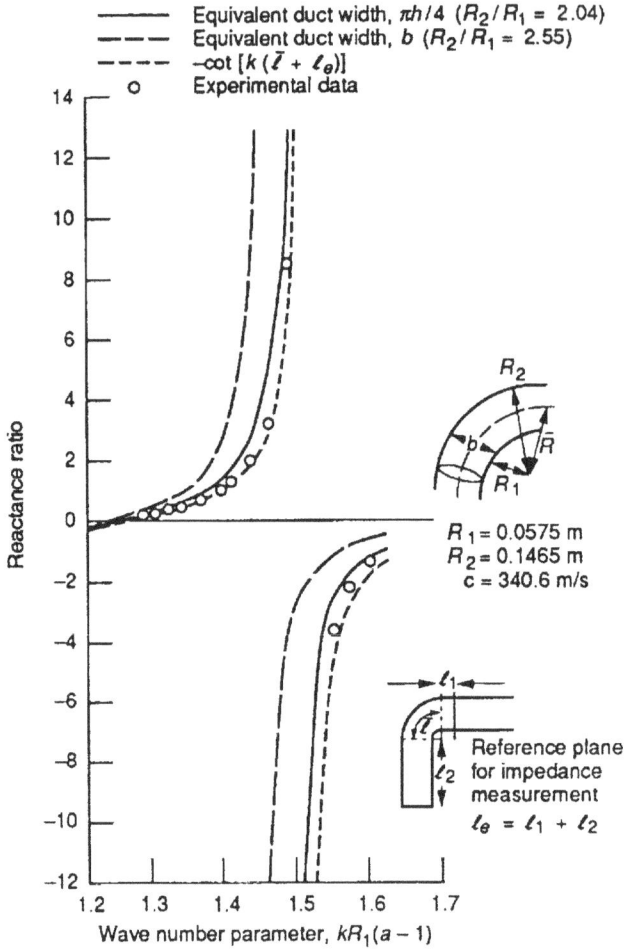

Figure 14.3.—Reactance ratio for circular-section curved bend. From Cummings (1974).

Because of its importance and uniqueness the analysis of toroidal ducts done by Keefe and Benade (1983) is given here.

Given this assumption, the total flow in the circular duct is the sum of the flows in the individual slices. Figure [14.4] shows that the cross-sectional area of a particular rectangular slice, located a distance z from the center of the circular cross section, is $(r_2^2 - z^2)^{1/2} \, dz$. The ratio a_p of the outer to inner radius of this slice may be written in terms of the bend parameter $B[=r_2/R_m=(a-1)/(a+1)]$.

$$a_p = \frac{1 + B[1 - (z/r_2)^2]^{1/2}}{1 - B[1 - (z/r_2)^2]^{1/2}} .$$

In order to determine the wave impedance and phase velocity we again use the transmission line formulation. The circular cross section arises as a superposition of all the slices, so that the transmission line representing the curved pipe is composed of the parallel combination of inertances, each shunted by its own compliance. The inertance L_i and compliance C_i per unit length of the ith rectangular slice are found...to be

$$L_i = \left(\rho/S\right)\left(v_0/kR_m\right)^2 = \rho/(R_m \ln a_p \, dz),$$

$$C_i = \left(S/\rho c^2\right),$$

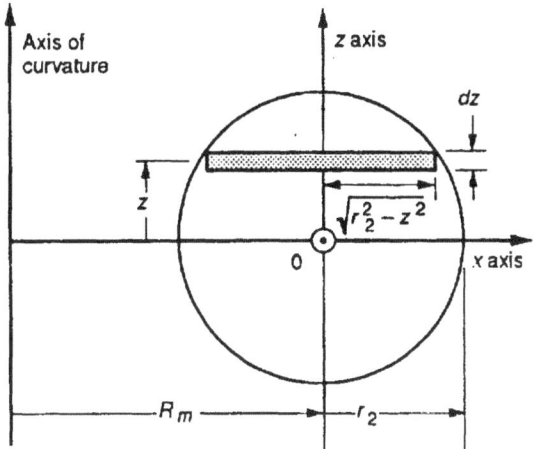

Figure 14.4.—Circular cross section of radius r_2 of toroidal bend, where for a given rectangular slice at a fixed height z the flow is assumed to be similar to that in a rectangular bend of height dz. From Keefe and Benade (1983).

where

$$S = (R_2 - R_1)dz.$$

The inertance L_t and the compliance C_t of the toroidal duct are obtained by summing over all the slices in parallel:

$$\frac{1}{L_t} = \sum_i \left(\frac{1}{L_i} \right),$$

$$C_t = \sum_i C_i .$$

The integrals obtained are

$$\frac{1}{L_t} = \left(\frac{R_m}{\rho} \right) \int_{-r_2}^{r_2} dz \ln a_p ,$$

$$C_t = (\pi r_2^2 / \rho c^2) = C_0 .$$

The compliance per unit length of a curved duct is simply the cross-sectional area divided by the bulk modulus. We wish to call particular attention here to the fact that the compliance of a toroid or rectangular bend is equal to the compliance C_0 of a straight pipe of the same cross section. In the low-frequency regime, the inertance is *always lowered* relative to its straight pipe value. The inertance of the toroidal pipe may be written in terms of the inertance per unit length L_0 of a straight pipe of similar cross section as follows:

$$L_t = L_0 \left(\pi B / 2I \right) ,$$

$$L_0 = \left(\rho / \pi r_2^2 \right) ,$$

$$I = \int_0^{\pi/2} \cos \theta \ln \left(\frac{1 + B \cos \theta}{1 - B \cos \theta} \right) d\theta,$$

where α is defined as arc $\sin\left(z/r_2\right)$.

The wave admittance Y and phase velocity v_p of the toroidal pipe are

$$Y = \left(C_t/L_t\right)^{1/2} = Y_0 \left(2I/\pi B\right)^{1/2},$$

$$v_p = (L_t C_t)^{-1/2} = c \left(2I/\pi B\right)^{1/2},$$

where

$$Y_0 = \left(C_0/L_0\right)^{1/2} = \left(\pi r_2^2/\rho c\right).$$

The integral I...has been numerically integrated by use of Simpson's rule; the relative shifts in wave admittance and phase velocity defined in the equation below are plotted in Fig. [14.5] as a function of the bend parameter B:

$$\frac{\Delta v_p}{c} = \frac{\Delta Y}{Y_0} = \left(\frac{2I}{\pi B}\right)^{1/2} - 1.$$

The figure shows that the shifts in admittance and phase velocity vary from 0% to 15% as the curvature of the bend increases, i.e., as B approaches unity....

We have measured the wave impedance and phase velocity of acoustical wave propagation in the curved section of a pipe of circular cross section, by means of normal mode frequency measurements on several combinations of straight and curved ducts....

One set of five semicircular curved pipe segments is assembled as a continuous helix [Fig. 14.6(a)]. Another set is assembled as a sinuous pipe of alternating curvature [Fig. 14.6(b)]...

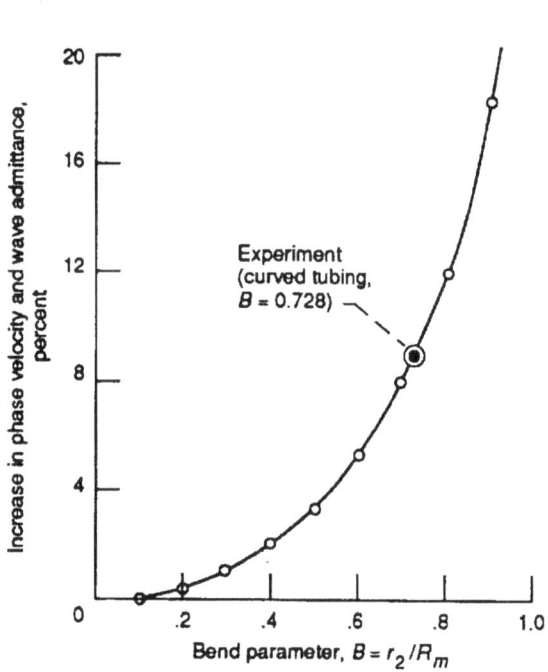

Figure 14.5.—Predicted shifts in wave admittance and phase velocity for wave in toroidal duct as function of bend parameter $B = r_2/R_m$, where r_2 is radius of the curved duct cross section and R_m is its radius of curvature. From Keefe and Benade (1983).

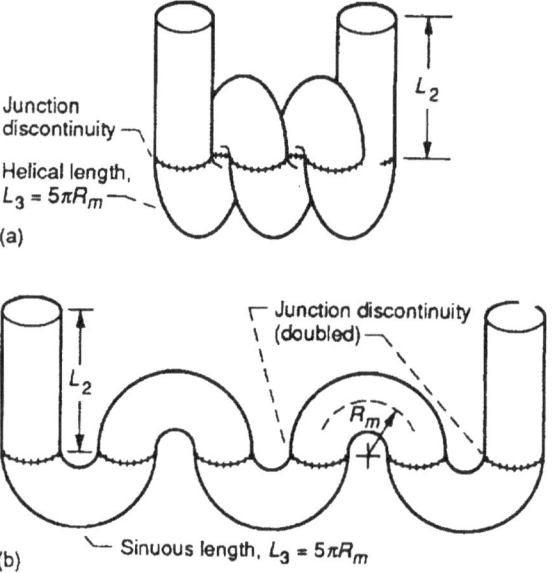

(a) Five curved segments joined as continuous helical spiral to form toroidal duct with 2½ turns.

(b) Five segments joined as sinuous duct so that adjacent segments have alternating curvature.

Figure 14.6.—Five semicircular segments from baritone horn tuning slide, each of whose midline lengths is $R_m = 12.7$ mm. Radius of circular cross section, $r_2 = 9.25$ mm. From Keefe and Benade (1983).

Our experimental result is that the wave-impedance, wave-admittance ($\Delta Y/Y_0 = -\Delta Z/Z_0$), and phase-velocity shifts are given by

$$\Delta Z/Z_0 = -6.3\%,$$

$$\Delta Y/Y_0 = 6.3\%,$$

$$\Delta v_p/c = 4.7\%.$$

It is notable that the measured wave-admittance shift is significantly larger than the phase-velocity shift, in disagreement with the earlier theoretical predictions that they should be equal. The discontinuity impedance Z_c as defined above is

$$Z_c = i(k\delta)Z_0 ,$$

[where δ is the length of a short segment of tubing] so that

$$Z_c/Z_0 \leq 0.1\%.$$

We see that the effect of the evanescent modes is very small compared to the propagating mode, since the discontinuity impedance is much less in magnitude than the propagating mode impedance.... The assumption made that the evanescent modes response is negligible relative to that of propagating mode is therefore shown to be valid....

The theory of wave propagation in the long-wavelength limit within strongly curved ducts of both rectangular and circular cross section has been investigated. The phase velocity and wave admittance are both increased in a curved duct relative to their straight duct values. Our experiments show that the wave-admittance shift is larger than the phase-velocity shift, although all theories predict that they should be equal. The observed shifts are smaller than the predicted shifts, which suggests that large shearing losses in the bulk of the fluid in the curved duct may be present.

The unresolved issues could be clarified by further low-frequency experiments using curved ducts of both rectangular and circular cross section.

Sound propagation in bends in a slender three-dimensional tubing of arbitrary (nonspecified) cross section was evaluated analytically by Ting and Miksis (1983) for four scalings of the wave number k. The slenderness of the tubing allowed them to bypass the difficulty of solving nonseparable toroidal coordinate systems by using the perturbation method (solving for motion inside the bent tubing). Slenderness is described by the perturbation parameter defined as the ratio $r_2/\ell \ll 1$, where r_2 is the reference length of the cross sections (here the radius of the circular tubing) and ℓ is the length of the bend or its radius of curvature. This perturbation analysis makes it possible to split the three-dimensional problem into two problems (a two-dimensional problem in a cross section and a one-dimensional problem along the curved centerline). Ting and Miksis concluded that for extremely long waves in curved tubing, sound propagation does not differ from that in straight tubing. For frequencies one order of magnitude higher, their results indicate that the phase of motion depends only on the length of the bend along its centerline and that the wave amplitude is inversely proportional to the square root of the tubing cross-sectional area.

For still higher frequencies (wavelengths of the order of the tubing radius) several modes can propagate and their number is governed by the tubing cross-sectional area. Some of the eigenvalues become imaginary, indicating evanescent waves. For shorter waves the propagation becomes a function of the bend's curvature.

An unusual study of modal wave propagation in bent elliptical piping was published by Furnell and Bies (1989). They described the way they solved the problem as follows:

In this paper, an approach similar to that proposed by Rice [1948] is used to facilitate a theoretical investigation of the modal nature of acoustic wave propagation within straight and circularly curved waveguides.... The problem is made amenable to an analysis [by] incorporating the theory of matrices [and] by seeking approximate mode solutions in the form of finite functional series expansions. By requiring that these solutions satisfy variational statements equivalent to the boundary value problem posed by [the Neumann condition] the series coefficients are determined via the Rayleigh-Ritz method....

...The interaction of the mode solutions at the junctions between sections of straight and curved waveguide are considered. This leads to the derivation of transmission and reflection matrices which relate the complex amplitudes of modes in a wave incident upon a section of curved waveguide, to those in waves transmitted through and reflected from it.... The derivation of these matrices utilizes techniques similar to those appearing in the papers by Tam [1976], Firth and Fahy [1984], Rostafinski [1976] and Fuller and Bies [1978a], and, once calculated, can be used to analyze complicated waveguide systems containing many curved and straight sections....

Finally,...the preceding theory is applied to a waveguide of elliptic cross-section, and results are presented which verify the feasibility of the proposed numerical algorithms.

The eigenfunctions and eigenvalues of a propagation were obtained by the classical separation of variables method, but the solution of the transverse field within the pipe was obtained by an approximate method based on the calculus of variations. Determination of the transmission and reflection characteristics of curved elliptical tubing relied on a system of matrices with eigenvectors and vectors containing complex amplitudes of the wave's eigenmodes. Next, to avoid using the complicated Mathieu functions in evaluating the characteristics of elliptical cross sections, Furnell and Bies (1989) used a transformation of coordinates that greatly simplified the analysis. Figure 14.7 shows contours of modes propagating in straight and curved elliptical tubing.

In an extensive study Firth and Fahy (1984) used series expansions to evaluate acoustic torus modes. Their Helmholtz equation in toroidal coordinates, in terms of the velocity potential, is

$$\frac{\partial^2 \psi}{\partial r^2} + \left(\frac{1}{r} - \frac{\cos \alpha}{R - r \cos \alpha} \right) \frac{\partial \psi}{\partial r} + \frac{1}{r^2} \frac{\partial^2 \psi}{\partial \alpha^2} + \frac{\sin \alpha}{r(R - r \cos \alpha)} \frac{\partial \psi}{\partial \alpha} + \frac{1}{(R - r \cos \alpha)^2} \frac{\partial^2 \psi}{\partial \theta^2} + k^2 \psi = 0$$

The assumed solution involves two potentials: one dependent on propagation within the pipe (pipe radius r and angular coordinate α in the pipe cross section are variables), the other involves the angular coordinate of the bend. Since $\psi(r,\alpha)$ is not separable in r and α, Firth and Fahy used a series solution to bypass the difficulty:

$$\psi_{r\alpha}(r,\alpha) = \sum_{n=0}^{\infty} A_n(r) \cos n\alpha + B_n(r) \sin n\alpha,$$

where $A_n(r)$ and $B_n(r)$ are functions representing the radial dependence, which will be different for each n.

They warn that a separate solution will be required for every frequency under consideration—a minor inconvenience. Radial dependence inside each cross section is represented by functions $A_n(r)$ and $B_n(r)$. For plane waves incident on the bend, they write

$$A_n(r) = \sum_{m=0}^{\infty} a_n^m J_n(X_n^m r/r_2),$$

where the a_n^m are constant coefficients and J_n is the Bessel function of the first kind of order n. The X_n^m are the zeros of the gradient of the function $J_n(x)$ and [r_2] is the radius of the cross section

that is, the sum of the radial functions of the modes in a straight pipe, which yields, as needed, zero normal gradient on the pipe wall. They give the results of a numerical analysis that includes matching solutions in the pipe bend with solutions in a straight pipe as shown in table 14.1.

Firth and Fahy (1984) performed calculations for a bend with $a = 2.33$ and $r_2/R_m = 0.4$ (where r_2 is the pipe radius) and for several modes of motion. In article notation, frequency is given as f/f_1, where f_1 is the first cutoff frequency for the cylindrical pipe. Angles α are angular locations in the pipe's cross sections measured from the horizontal. Figure 14.8 gives the calculated radial and angular distributions of normalized acoustical pressure. The modes, of course, resemble those propagating in a straight pipe and, as frequency

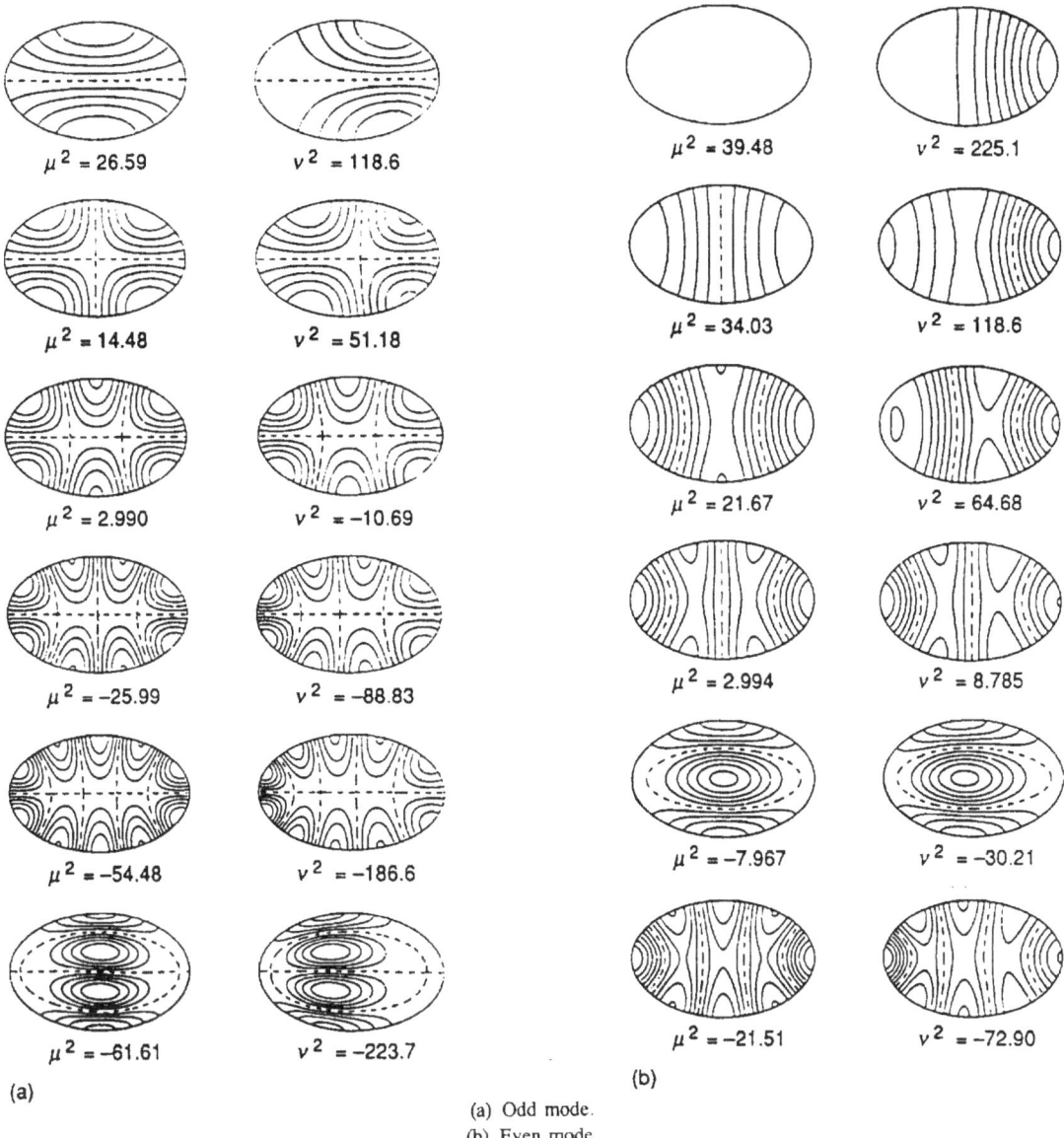

$\mu^2 = 26.59$ $\nu^2 = 118.6$ $\mu^2 = 39.48$ $\nu^2 = 225.1$

$\mu^2 = 14.48$ $\nu^2 = 51.18$ $\mu^2 = 34.03$ $\nu^2 = 118.6$

$\mu^2 = 2.990$ $\nu^2 = -10.69$ $\mu^2 = 21.67$ $\nu^2 = 64.68$

$\mu^2 = -25.99$ $\nu^2 = -88.83$ $\mu^2 = 2.994$ $\nu^2 = 8.785$

$\mu^2 = -54.48$ $\nu^2 = -186.6$ $\mu^2 = -7.967$ $\nu^2 = -30.21$

$\mu^2 = -61.61$ $\nu^2 = -223.7$ $\mu^2 = -21.51$ $\nu^2 = -72.90$

(a)   (b)

(a) Odd mode.
(b) Even mode.

Figure 14.7.—Contours in straight and curved waveguides, where μ and ν are eigenvalues in straight and bent elliptical ducts, respectively. From Furnell and Bies (1989).

TABLE 14.1.—ORDER OF
CYLINDER MODES

[From Firth and Fahy (1984).]

Mode	Order, n	Terms of series, m	X_n^m
1	0	0	0
2	1	1	1.84118
3	2		3.05424
4	0		3.83171
5	3		4.20119
6	4		5.31755
7	1	2	5.33144

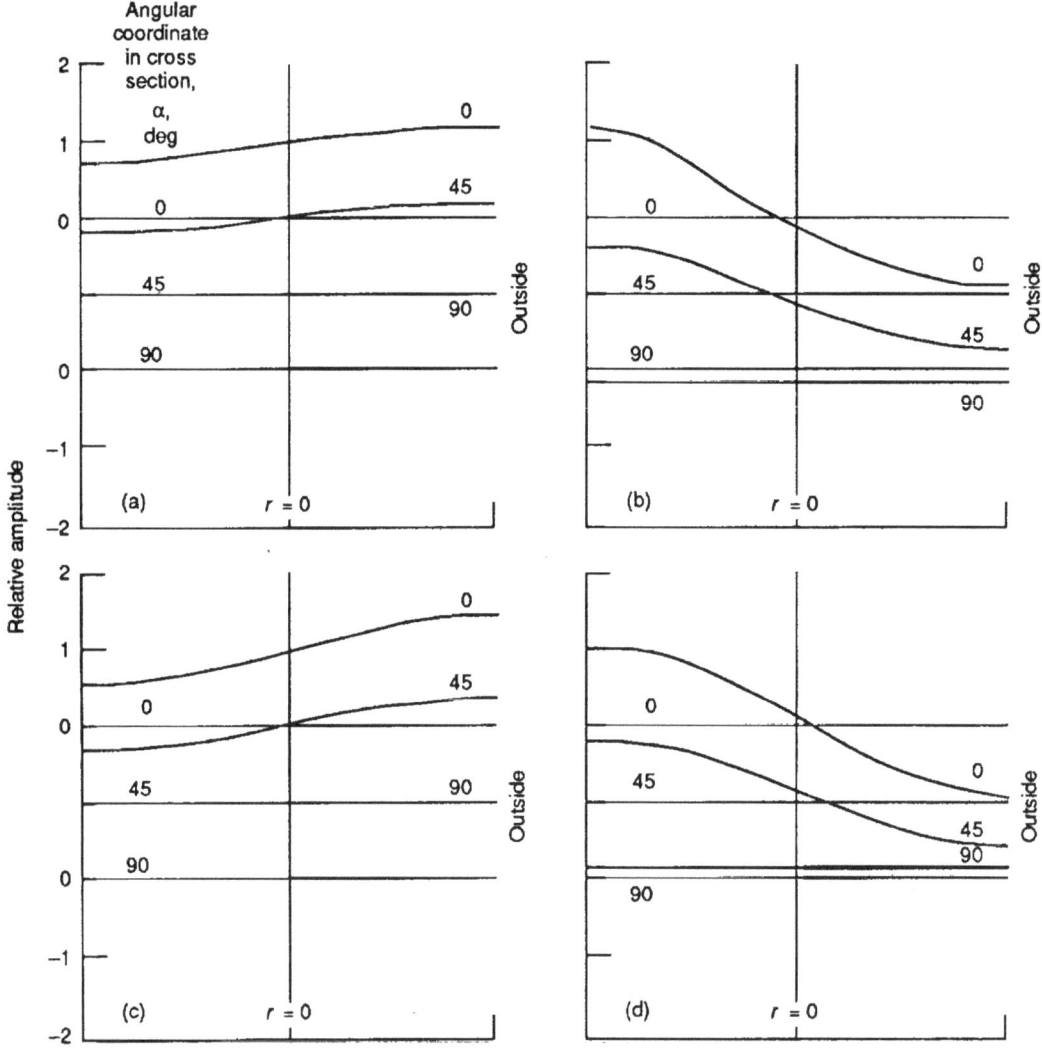

(a) Mode 1*; $\nu = 2.785$; $r_2/R_m = 0.4$; $f/f_1 = 0.6$, where f_1 is the first cutoff frequency.

(b) Mode 2; $\nu = 3.41i$; $r_2/R_m = 0.4$, $f/f_1 = 0.6$.

(c) Mode 1; $\nu = 11.294$; $r_2/R_m = 0.2$, $f/f_1 = 1.2$.

(d) Mode 2*; $\nu = 5.83$; $r_2/R_m = 0.2$, $f/f_1 = 1.2$.

Figure 14.8.—Symmetric acoustic torus modes. From Firth and Fahy (1984).

decreases, the modes become more like the cylindrical modes. The calculated torus wave numbers for several modes are shown in figures 14.9 and 14.10. The second graph includes data calculated by Cummings (1974) for a bent duct. The transmission and reflection coefficients have been calculated for a 90° bend connected to two straight pipes. Figure 14.11 gives the amplitude and phase of the incident waves (three modes) at the inlet junction as a function of frequency up to twice the first-mode frequency in a straight pipe. The basic plane-wave mode is transmitted well.

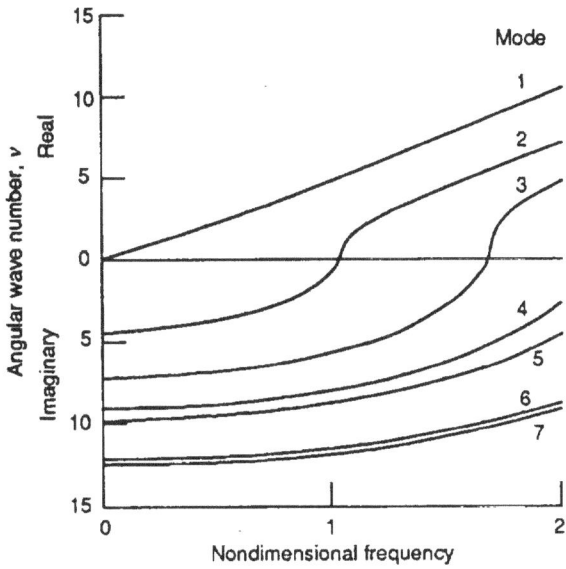

Figure 14.9.—Dispersion curves for torus modes for radius ratio, $r_2/R_m = 0.4$. From Firth and Fahy (1984).

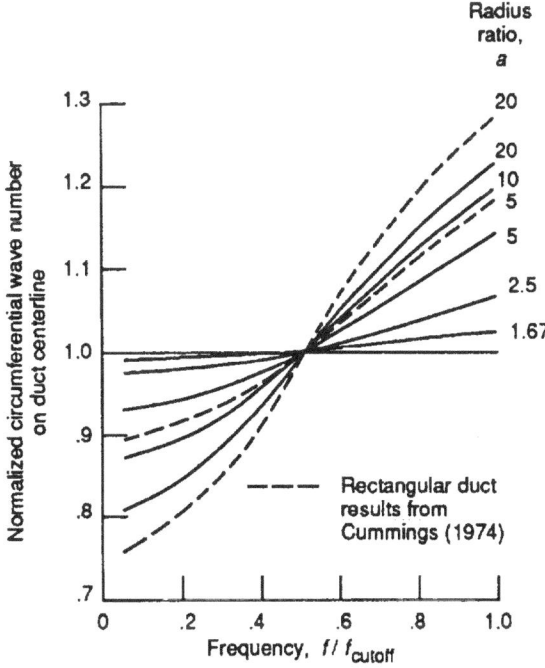

Figure 14.10.—Variation of bend wave number with frequency for various bend radii. From Firth and Fahy (1984).

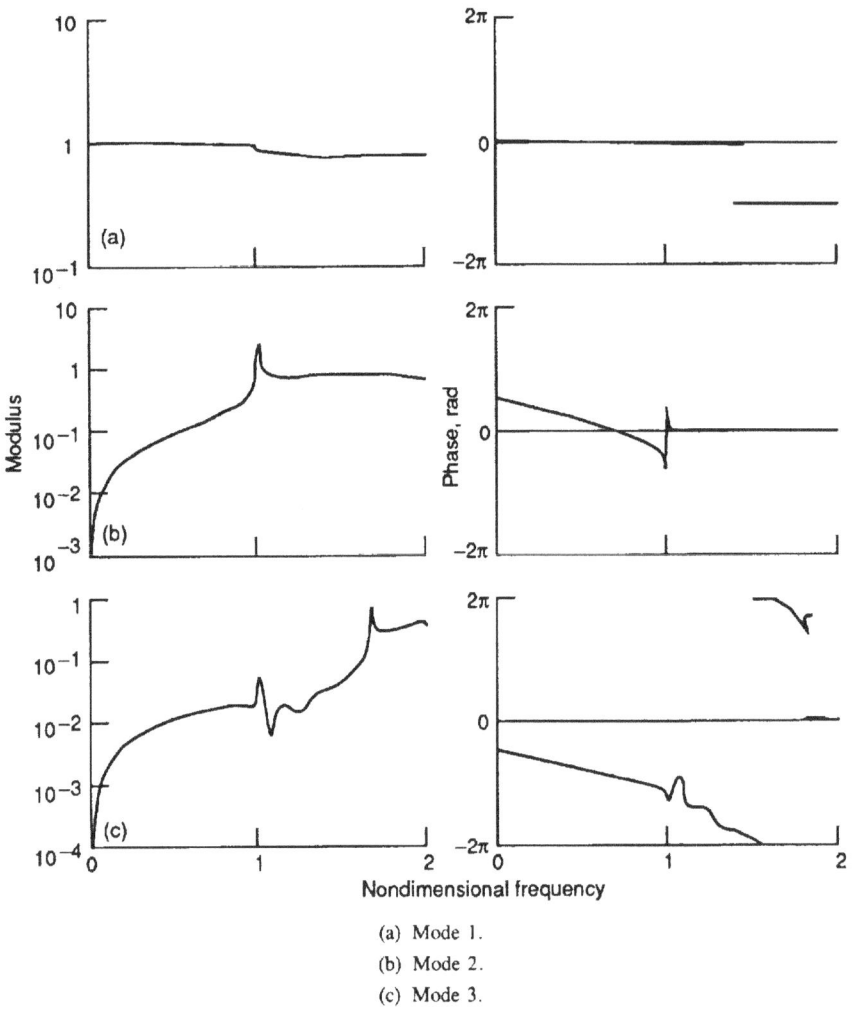

(a) Mode 1.

(b) Mode 2.

(c) Mode 3.

Figure 14.11.—Reflection and transmission of acoustic modes at the first junction of a circular bend. Radius ratio, $r_2/R_m = 0.4$; coefficient, B^+; incident plane wave. From Firth and Fahy (1984).

15.0 Sound Propagation in Rounded Corners

Although not exactly belonging to the study of acoustical behavior in cylindrical bends, the paper by Cabelli and Shepherd (1981) on acoustical characteristics of rounded corners (see fig. 15.1) merits mention in this monograph because it further illustrates the effects of rounded corners, axial partition in a bend, and selective, frequency-dependent reflections. Out of the experimental data reported in their paper, one figure may be of special interest here (fig. 15.2). It illustrates the influence of the inner and outer radii on the acoustic energy reflected in a 90° mitered bend with rounded corners at wave number parameter $kR_1(a - 1) = 4.5$. They comment as follows:

> Clearly a radius on the outer wall of the bend greatly reduces reflection while a radiused inner wall increases reflection, particularly where an outer radius exists. Therefore, a generous inner wall radius is beneficial acoustically and aerodynamically (at most frequencies between cut-on of the first and second cross modes) whilst a curved outer wall represents a trade off between aerodynamic and acoustic performance.

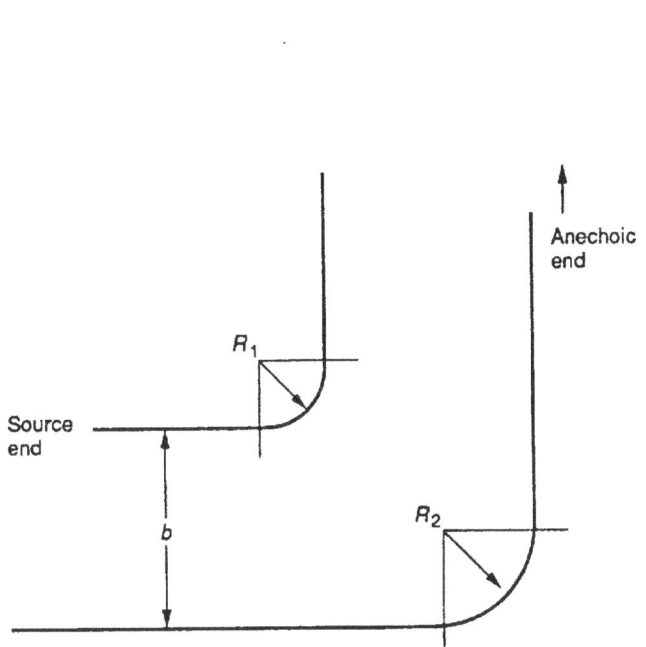

Figure 15.1.—Rounded corner geometry. From Cabelli and Shepherd (1981).

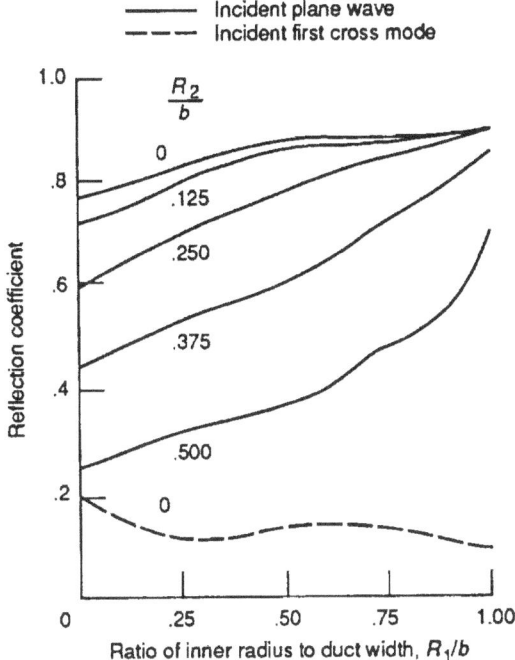

Figure 15.2.—Variation of reflection coefficient with bend radii for 90° bend. Wave number parameter, $kR_1(a - 1) = 4.5$. From Cabelli and Shepherd (1981).

16.0 Main Acoustical Characteristics of Bends

Work done so far has contributed considerably to the knowledge of sound propagation in bends. In early papers published before the 1970's at least one new characteristic, the angular wave number, was defined. Krasnushkin (1945) coined the term "angular wave number," a nondimensional parameter, as opposed to the dimensional wave number, which characterizes motion in straight conduits. In the 1970's and 1980's a wealth of information was obtained. The main features of sound propagation in curved ducts and pipes are as follows:

(1) The existence of a nondimensional propagation parameter called the angular wave number, which is the order of the Bessel function, the eigenvalue of the characteristic function of motion. Angular wave number is a function of both the wave number parameter kR_1, which is the argument of the Bessel functions, and of the bend radius ratio $a = R_2/R_1$, which also defines the characteristic equation. In other words, the radial geometry of the bend and the frequency of the imposed sound determine this parameter. In spite of the importance of the angular wave number, the bend's acoustical characteristics can be established by numerical methods that do not require knowledge of the angular wave number.

(2) Similarity between motion in pipes and ducts. Propagation in pipes does not differ to any radical degree from motion in ducts so that data available from studies of rectangular bends constitute a satisfactory approximation for engineering purposes.

(3) Inability of the plane wave (0,0) mode to propagate in a bend. A plane wave moving in a straight duct changes its characteristics when it enters a bend. Particle velocities are no longer uniform, and even in the low frequency range the pressure distribution is not exactly uniform. In extremely narrow ducting, however, long waves propagate as if the duct were straight. On the other hand, sound propagation in bends is always characterized by the appearance of radial vibrations.

(4) The existence of many modes. As in a straight duct many different modes can propagate in bends; also a great number of evanescent waves are generated at every junction between a straight and a curved section of a ducting system. They attenuate rapidly but go farther in sharp bends (large a) than in narrow ducting.

(5) Dependence of phase velocity in bends on frequency. At frequencies lower than $kR_1(a - 1)/\pi = 0.5$ a wave moves faster; at higher frequencies it moves slower. At frequencies corresponding to exactly 0.5, waves propagate with phase velocity entirely independent of a (i.e., as if there were no duct curvature).

(6) Resonances in bends. As in straight ducts, resonances appear in bends, but the curvature introduces a shift in the resonant frequencies.

(7) Uncertainty about reflections. The matter of reflections in hard-wall bends is not clear as yet. Some experiments indicate that the transmissivity of bends, even extremely sharp bends, is excellent for extremely low to low frequencies. At higher frequencies significant reflections have been measured.

(8) Lower impedance than straight ducts. The impedance of hard-wall ducts is slightly less than the impedance of corresponding (lengthwise along the centerline and the mean radius $(R_1 + R_2)/2$) straight ducts.

(9) Effectiveness of acoustic lining. Acoustically lined bends, of all radius ratios but of equal median length, attenuate sound about as well as a corresponding straight duct. Not all modes are capable of propagating in a curved acoustically lined duct: only the fundamental mode can propagate at all frequencies.

(10) Effectiveness of turning vanes. Bends equipped with turning vanes (flat or shaped), if properly designed, may become excellent sound attenuators in a relatively wide range of frequencies.

17.0 Concluding Remarks

Many analytical and experimental studies of bends with many different degrees of sharpness and in a wide frequency range have determined the main characteristics of sound propagation in curved ducts. The data, as reported herein, and supplemented when needed by the original papers, should give satisfactory understanding of the penalties and benefits involved in using bends in acoustical systems.

However, several important areas require serious additional analytical and experimental effort:

(1) An overall encompassing code is needed for obtaining all the characteristics of sound propagation in any particular bend for any given frequency.

(2) The matter of reflections should be clarified and the reflectivity and transmissivity of both hard-wall and acoustically lined bends determined.

(3) The acoustical performance of cylindrical pipe bends should be studied in greater detail.

(4) The effect of flow on sound transmission in bends should be determined.

(5) The merits of non-locally-reactive acoustical linings in bends relative to standard, Helmholtz resonator linings should be investigated further.

Lewis Research Center
National Aeronautics and Space Administration
Cleveland, Ohio, August 2, 1990

Appendix A
Derivation of Roots for Long Waves in Hard-Wall Bends

There are no known tables* for the roots ν_m of the characteristic equation

$$[J'_{\nu_m}(kR_1)J'_{-\nu_m}(akR_1) - J'_{\nu_m}(akR_1)J'_{-\nu_m}(kR_1)] = 0$$

In order to evaluate the ν_m's, the J'_ν and $J'_{-\nu}$ are expanded by increasing the powers of the arguments akR_1 and kR_1. Limiting the expansion in the first approximation to the first term, which yields satisfactory results when arguments of the Bessel functions are much less than 1, gives

$$J'_\nu(kr) \simeq \frac{\nu}{2^\nu \Gamma(\nu + 1)}(kr)^{\nu - 1} + \ldots$$

$$J'_{-\nu}(kr) \simeq \frac{\nu}{2^{-\nu}(1 - \nu)}(kr)^{-\nu - 1} + \ldots$$

and using a recurrence relation for the gamma function gives, after substitutions,

$$\frac{\nu^2}{\sin \pi\nu \Gamma(\nu + 1)(1 - \nu)} \left[(kR_1)^{-\nu - 1}(akR_1)^{\nu - 1} - (kR_1)^{\nu - 1}(akR_1)^{-\nu - 1} \right] = 0$$

and finally

$$\frac{\nu}{\pi(kR_1)^2}(a^{\nu - 1} - a^{-\nu - 1}) = 0$$

Solution $\nu = 0$ must be rejected as a general solution. Therefore, the general solution must satisfy the equation

$$a^{\nu - 1} = a^{-\nu - 1}$$

which may be put in the form

$$a^{2\nu} = 1 \qquad \text{or} \qquad e^{2\nu \ln a} = 1$$

Hence, $2\nu \ln a = 2m\pi i$, that is,

$$\nu_m = \frac{m\pi i}{\ln a} \qquad m = 1, 2, 3, \ldots \tag{A1}$$

Better approximations will be given by the second and following terms of the expansion of J'_ν and $J'_{-\nu}$. After algebraic manipulations and dropping the term containing $(kR_1)^2$,

$$\frac{\sin \pi\nu}{16\pi\nu(1 + \nu)(1 - \nu)} \left[\nu(1 - \nu)(2 + \nu)(a^{-\nu - 1} - a^{1 + \nu}) + (1 + \nu)(2 - \nu)\nu(a^{\nu - 1} - a^{1 - \nu}) \right.$$
$$\left. + 4\nu^2(1 - \nu)(1 + \nu)(kR_1)^{-2}(a^{\nu - 1} - a^{-\nu - 1}) \right] = 0$$

---

*A table of roots for $\nu_m = 0.5, 1.5, 2.5, \ldots$ has been published by Rostafinski (1974c). In the present case (extremely small ν_m) it is necessary to extrapolate the tabulated data.

In order to improve on the first approximation (eq. (A1)), it is assumed that

$$\nu_m = \frac{m\pi i}{\ln a} + \frac{\ln(1 + \epsilon_m)}{\ln a}$$ (A2)

where ϵ_m is a small quantity. Solving and remembering that

$$\ln z = \ln |z| + i(\arg z + 2n\pi) \qquad n = 0, \pm 1, \ldots$$

and substituting again, since

$$\frac{1}{1 + \epsilon} \cong 1 - \epsilon$$

gives

$$(2 - \nu^2)(a^{-1} - a^1) + 2\nu(1 - \nu^2)(kR_1)^{-2}\left(\frac{2\epsilon}{a}\right) = 0$$

$$\epsilon_m = -\frac{(2 - \nu^2)(kR_1)^2}{4\nu(1 - \nu^2)}(1 - a^2) = -i\,\frac{(kR_1)^2(a^2 - 1)\left[2 + \dfrac{m^2\pi^2}{(\ln a)^2}\right]}{4m\pi\left[1 + \dfrac{m^2\pi^2}{(\ln a)^2}\right]}\ln a$$

Finally,

$$\nu_m = i\left\{\frac{m\pi}{\ln a} - \frac{(kR_1)^2(a^2 - 1)\left[2 + \dfrac{m^2\pi^2}{(\ln a)^2}\right]}{4m\pi\left[1 + \dfrac{m^2\pi^2}{(\ln a)^2}\right]}\right\} \qquad m = 1, 2, 3, \ldots$$

Now for $m = 0$ in equation (A2)

$$\nu_0 \simeq \frac{\ln(1 + \epsilon_0)}{\ln a}$$

Substituting again and neglecting terms in ν_0^2 gives

$$\left[2 - \frac{\ln(1 + \epsilon_0)}{\ln a}\right]\frac{1}{a(1 + \epsilon_0)} - a(1 + \epsilon_0) + \left[2 + \frac{\ln(1 + \epsilon_0)}{\ln a}\right]\left(\frac{1 + \epsilon_0}{a} - \frac{a}{1 + \epsilon_0}\right)$$

$$+ 4\frac{\ln(1 + \epsilon_0)}{\ln a}(kR_1)^{-2}\left[\frac{1 + \epsilon_0}{a} - \frac{a}{a(1 + \epsilon_0)}\right] = 0$$

84

Solving gives

$$\left[2 - \frac{\ln(1 + \epsilon_0)}{\ln a} \right] \frac{1 - \epsilon_0}{a} - a(1 + \epsilon) + \left[2 + \frac{\ln(1 + \epsilon_0)}{\ln a^{\cdot}} \right] \frac{1 + \epsilon_0}{a} - a(1 - \epsilon)$$

$$+ 4 \frac{\ln(1 + \epsilon_0)}{\ln a} (kR_1)^{-2} \left(\frac{2\epsilon_0}{a} \right) = 0$$

$$-4 \left(a - \frac{1}{a} \right) + 2 \frac{\ln(1 + \epsilon_0)}{\ln a} \epsilon_0 \left(a + \frac{1}{2} \right) + \frac{8\epsilon_0}{a} \frac{\ln(1 + \epsilon_0)}{\ln a} (kR_1)^{-2} = 0$$

Finally,

$$\frac{\epsilon_0 \ln(1 + \epsilon_0)}{\ln a} = \frac{2\left(a - \frac{1}{2} \right)}{\frac{4}{a}(kR_1)^{-2} + a + \frac{1}{2}}$$

Since $\ln(1 + \epsilon_0) \simeq \epsilon_0$,

$$\frac{\epsilon_0^2}{\ln a} = \frac{2(a^2 - 1)}{4(kR_1)^{-2} + a^2 + 1}$$

and

$$v_0^2 = \frac{\dfrac{2(a^2 - 1)}{\ln a}}{4(kR_1)^{-2} + a^2 + 1}$$

Since the result depends on v_0^2, the preceding steps must be retraced and evaluation performed without neglecting terms in v_0^2. After algebraic manipulations a new expression for v_0 is obtained:

$$v_0 = \left[\frac{\dfrac{2(a^2 - 1)}{\ln a}}{4(kR_1)^{-2} + a^2 + 1 + \dfrac{a^2 - 1}{\ln a}} \right]^{1/2}$$

There is an infinite set of pure imaginary roots $v_m = i(m\pi/\ln a)$, $m = 1, 2, 3, \ldots$ and one single real root v_0. The uniqueness of the obtained real root can be verified by substituting for a^v the power series

$$a^v = 1 + v \ln a + \frac{v^2(\ln a)^2}{2} \cdots$$

In the first approximation,

$$\nu_0 = \left[\frac{\dfrac{2(a^2 - 1)}{\ln a}}{4(kR_1)^{-2}} \right]^{1/2} = (kR_1) \left(\frac{a^2 - 1}{2 \ln a} \right)^{1/2}$$

This result was obtained with only the first two terms of the series for a^{ν}. When three terms of this series were used and when small terms of the fourth order were neglected, the result was

$$\nu_0 = \left[\frac{\dfrac{2(a^2 - 1)}{\ln a}}{4(kR_1)^{-2} + a^2 + 1 + \dfrac{a^2 - 1}{\ln a} - (a^2 - 1)\ln a} \right]^{1/2}$$

The equation for ν_0 was thus verified by series expansion, and the uniqueness of the root ν_0 was established.

Appendix B
Propagating-Mode and Evanescent-Wave Solutions of Characteristic Equation for Long Waves in Hard-Wall Bends

Let us consider the characteristic function

$$F_{\nu_m}(r) = J'_{\nu_m}(kR_1) J_{-\nu_m}(kr) - J_{\nu_m}(kr) J'_{-\nu_m}(kR_1) \qquad \text{where } m = 0, 1, 2, \ldots$$

Real Root Solution

Let us consider the solution of F_{ν_0} pertaining to the real root ν_0. Using the series expansions for $J_{\nu_0}(kr)$ and $J_{-\nu_0}(kR_1)$ and for $J'_{\nu_0}(kr)$ and $J'_{-\nu_0}(kR_1)$ and equations for the gamma functions and rearranging those expressions gives

$$F_{\nu_0}(r,R_1) = \frac{\sin(\pi\nu_0)}{4(kR_1)\nu_0\pi(1 - \nu_0^2)} \left\{ \left(\frac{r}{R_1}\right)^{\nu_0} \left[-4\nu_0 + 4\nu_0^3 - 2(kR_1)^2 + \nu_0^2(kR_1)^2 - \nu_0(kR_1)^2 \right] \right.$$

$$+ \left(\frac{r}{R_1}\right)^{\nu_0} (kr)^2 \left[\frac{\nu_0 - \nu_0^2 + (kR_1)^2}{2} - \frac{\nu_0(kR_1)^2}{4} \right] + \left(\frac{r}{R_1}\right)^{-\nu_0} \left[-4\nu_0 + 4\nu_0^3 \right.$$

$$\left. + 2(kR_1)^2 - \nu_0^2(kR_1)^2 - \nu_0(kR_1)^2 \right] + \left(\frac{r}{R_1}\right)^{-\nu_0} (kr)^2 \left[\frac{\nu_0 + \nu_0^2 - (kR_1)^2}{2} - \frac{\nu_0(kR_1)^2}{4} \right] \right\}$$

Next, using power series for the exponential functions

$$\left(\frac{r}{R_1}\right)^{\pm\nu_0} = 1 \pm \nu_0 \ln\frac{r}{R_1} + \frac{\nu_0^2 \left(\ln\frac{r}{R_1}\right)^2}{2!} \pm \frac{\nu_0^3 \left(\ln\frac{r}{R_1}\right)^3}{3!} + \ldots$$

and neglecting terms in ν_0^4 and terms containing products $(kr)^2 (kR_1)^2$ that are of the same order, we get

$$F_{\nu_0}(r,R_1) = \frac{\sin(\pi\nu_0)}{4\pi(kR_1)(1 - \nu_0^2)} \left\{ \left[-8 + 8\nu_0^2 - 2(kR_1)^2 - 4(kR_1)^2 \ln\frac{r}{R_1} + 2\nu_0^2(kR_1)^2 \ln\frac{r}{R_1} \right. \right.$$

$$\left. - 4\nu_0^2 \left(\ln\frac{r}{R_1}\right)^2 - \nu_0^2(kR_1)^2\left(\ln\frac{r}{R_1}\right)^2 - \frac{2}{3}\nu_0^2(kR_1)^2\left(\ln\frac{r}{R_1}\right)^3 \right]$$

$$\left. + (kr)^2 \left[2 - 2\nu_0^2 \ln\frac{r}{R_1} + \nu_0^2\left(\ln\frac{r}{R_1}\right)^2 \right] \right\}$$

This general solution verifies the differential equation for a wide range of radius ratios $a = R_2/R_1$. For $4 < a < 1$ the error is negligible; for $a = 10$ the error is approximately 1 percent.

This equation, if greatly simplified by eliminating products of small terms becomes

$$F_{\nu_0}(r,R_1) = \frac{\sin(\pi\nu_0)}{4\pi(kR_1)} \left[-8 + 8\nu_0^2 - 2(kR_1)^2 - 4\nu_0^2\left(\ln\frac{r}{R_1}\right)^2 + 2(kr)^2 - 4(kR_1)^2 \ln\frac{r}{R_1} \right]$$

and, when applied to the differential equation, still yields a satisfactory expression for ν_0 at $r = R_2$, namely

$$\nu_0^2 \cong \frac{\dfrac{2(a^2 - a)}{\ln a}}{4(kR_1)^{-2}} = (kR_1)^2 \frac{a^2 - 1}{2 \ln a}$$

For numerical calculations F_{ν_0} is, with high accuracy,

$$F_{\nu_0} \cong \frac{2 \sin(\pi \nu_0)}{\pi(kR_1)}$$

Imaginary Root Solutions

Let us consider the Bessel functions of pure imaginary order and real argument and use tables of functions published by Buckens (1963). The very basic relations (from Buckens, 1963) are

$$F_\mu(kr) + iG_\mu(kr) = 2^{i\mu}\Gamma(1 + i\mu)J_{i\mu}(kr) \ldots$$

$$F_\mu(kr) - iG_\mu(kr) = 2^{-i\mu}\Gamma(1 - i\mu)J_{-i\mu}(kr)$$

The left sides of these equations are complex conjugates. As $\overline{\Gamma(1 + i\mu)} = \overline{\Gamma(1 + i\mu)}$, it can be concluded that $2^{i\mu}J_{-i\mu}(kr)$ is a complex conjugate of $2^{-i\mu}J_{i\mu}(kr)$. In these equations $i\mu = \nu$ and

$$G_\mu(kr) = A_\mu(kr) \sin[\mu \ln(kr)] + B_\mu(kr) \cos[\mu \ln(kr)]$$

$$F_\mu(kr) = A_\mu(kr) \cos[\mu \ln(kr)] - B_\mu(kr) \sin[\mu \ln(kr)]$$

The functions $A_\mu(kr)$ and $B_\mu(kr)$ are

$$A_\mu(kr) = \sum_{g=0}^{\infty} (\xi)_{2g}(ikr)^{2g}$$

$$B_\mu(kr) = \sum_{g=0}^{\infty} (\zeta)_{2g}(ikr)^{2g}$$

where $g = 0, 1, 2, 3, \ldots$, $(\xi)_0 = 1$, $(\zeta)_0 = 0$, and

$$\zeta_{2g} = \frac{g\zeta_{2g-2} - \mu\xi_{2g-2}}{4g(\mu^2 + g^2)}$$

$$\xi_{2g} = \frac{g\xi_{2g-2} + \mu\zeta_{2g-2}}{4g(\mu^2 + g^2)}$$

Neglecting the fourth and higher powers of kr gives

$$F_\mu(kr) = \left[1 - \frac{(kr)^2}{4(1 + \mu^2)}\right] \cos[\mu \ln(kr)] - \frac{(kr)^2}{4(1 + \mu^2)} \sin[\mu \ln(kr)]$$

$$G_\mu(kr) = \left[1 - \frac{(kr)^2}{4(1 + \mu^2)}\right] \sin[\mu \ln(kr)] + \frac{(kr)^2}{4(1 + \mu^2)} \cos[\mu \ln(kr)]$$

Substituting into the basic equations for $F_\mu(kr) \pm iG_\mu(kr)$ results in expressions for $J_{i\mu}(kr)$ and $J_{-i\mu}(kr)$ for $kr << 1$ (terms in $(kr)^2$ neglected).

$$\cos[\mu \ln(kr)] + i \sin[\mu \ln(kr)] = 2^{i\mu}\Gamma(1 + i\mu)J_{i\mu}(kr)$$

$$\cos[\mu \ln(kr)] - i \sin[\mu \ln(kr)] = 2^{i\mu}\Gamma(1 - i\mu)J_{-i\mu}(kr)$$

Taking the derivative of the basic equation for $F_\mu(kr) + iG_\mu(kr)$ with respect to the argument kr of the Bessel function gives

$$2^{i\mu}\Gamma(1 + i\mu)J'_{i\mu}(kr) = \sin[\mu \ln(kr)]\left[2\varsigma_2(kr) + \mu\xi_2(kr) - \frac{\mu}{kr}\right] + \cos[\mu \ln(kr)][\mu\varsigma_2(kr) - 2\xi_2(kr)]$$

$$+ i\left\{-\cos[\mu \ln(kr)]\left[2\varsigma_2(kr) + \mu\xi_2(kr) - \frac{\mu}{kr}\right] + \sin[\mu \ln(kr)][\mu\varsigma_2(kr) - 2\xi_2(kr)]\right\}$$

All terms containing $\varsigma_2(kr)$ and $\xi_2(kr)$ are small in relation to μ/kr and in the first approximation may be neglected. The resulting equations are

$$2^{i\mu}\Gamma(1 + i\mu)J'_{i\mu}(kr) = -\frac{\mu}{kr}\sin[\mu \ln(kr)] + i\frac{\mu}{kr}\cos[\mu \ln(kr)]$$

$$2^{-i\mu}\Gamma(1 - i\mu)J'_{-i\mu}(kr) = -\frac{\mu}{kr}\sin[\mu \ln(kr)] - i\frac{\mu}{kr}\cos[\mu \ln(kr)]$$

Therefore F_{ν_m} for $m \neq 0$ can be formed as follows:

$$F_{\nu_m} = -2i\frac{\mu_m}{kR_1}\frac{\sin[\mu_m \ln(kr)] \sin[\mu_m \ln(kR_1)] + \cos[\mu_m \ln(kr)] \cos[\mu_m \ln(kR_1)]}{\Gamma(1 + i\mu_m)\Gamma(1 - i\mu_m)}$$

Since

$$\Gamma(1 + i\mu_m)\Gamma(1 - i\mu_m) = \frac{\pi\mu_m}{\sinh(\pi\mu_m)}$$

using trigonometric transformation results in

$$F_{\nu_m} = i\frac{2 \sinh(\pi\mu_m)}{\pi(kR_1)}\cos\left(\mu_m \ln\frac{r}{R_1}\right)$$

This equation shows that F_{ν_m} is a pure imaginary number and is a function of radius r.

References

Baumeister, K.J.; and Rice, E.J. (1975) A Difference Theory for Noise Propagation in an Acoustically Lined Duct With Mean Flow. Aeroacoustics: Jet and Combustion Noise; Duct Acoustics, H.T. Nagamatsu et al., eds., Progress in Astronautics and Aeronautics, Vol. 37, AIAA, pp. 435–453.

Baumeister, K.J. (1989) Acoustic Propagation in Curved Ducts With Extended Reacting Wall Treatment. NASA TM–102110, 1989.

Buchholz, H. (1939) Der Einfluss der Kruemmung von rechteckigen Hohlleitern auf das Phasenmass ultrakurzer Wellen. Electrische Nachrichten-Technik, vol. 16, no. 3, Mar., pp. 73–85.

Buckens, F. (1963) *Tables of Bessel Functions of Imaginary Order*. Institut de Mecanique et Mathematiques Appliquees, Universite Catholique de Louvain, Belgium, Dec. (Avail. NTIS, AD–602755).

Cabelli, A. (1980) The Acousic Characteristics of Duct Bends. J. Sound Vib., vol. 68, no. 3, Feb. 8, pp. 369–388.

Cabelli, A.; and Shepherd, I.C. (1981) The Influence of Geometry on the Acoustic Characteristics of Duct Bends for Higher Order Modes. J. Sound Vib., vol. 78, no. 1, Sept. 8, pp. 119–129.

Cummings, A. (1974) Sound Transmission in Curved Duct Bends. J. Sound Vibr., vol. 35, no. 4, Aug. 22, pp. 451–477.

El-Raheb, M. (1980) Acoustic Propagation in Rigid Three-Dimensional Waveguides. J. Acous. Soc. Am., vol. 67, no. 6, June, pp. 1924–1930.

El-Raheb, M.; and Wagner, P. (1980) Acoustic Propagation in Rigid Sharp Bends and Branches. J. Acoust. Soc. Am., vol. 67, no. 6, June, pp. 1914–1923.

Firth, D.; and Fahy, F.J. (1984) Acoustic Characteristics of Circular Bends in Pipes. J. Sound Vibr., vol. 97, no. 2, Nov. 22, pp. 287–303.

Fuller, C.R.; and Bies, D.A. (1978a) Propagation of Sound in a Curved Bend Containing a Curved Axial Partition. J. Acoust. Soc. Am., vol. 63, no. 3, Mar., pp. 681–686.

Fuller, C.R.; and Bies, D.A. (1978b) A Reactive Acoustic Attenuator. J. Sound Vibr., vol. 56, no. 1, Jan. 8, pp. 45–59.

Furnell, G.D.; and Bies, D.A. (1989) Characteristics of Modal Wave Propagation Within Longitudinally Curved Acoustic Waveguides. J. Sound Vibr., vol. 130, no. 3, May 8, pp. 405–423.

Grigor'yan, F.E. (1969) Theory of Sound Wave Propagation in Curvilinear Waveguides. Akust. Zh., vol. 14, 1968, pp. 376–384. (English Translation: Sov. Phys. Acoust., vol. 14, no. 3, Jan.-Mar., pp. 315–321.)

Grigor'yan, F.E. (1970) Soundproofing by Means of Ducts With Curved Porous Walls. Sov. Phys. Acoust., vol. 16, no. 2, Oct.-Dec., pp. 192–196.

Keefe, D.H.; and Benade, A.H. (1983) Wave Propagation in Strongly Curved Ducts. J. Acoust. Soc. Am., vol. 74, no. 1, July, pp. 320–332.

Ko, S.-H. (1979) Three-Dimensional Acoustic Waves Propagating in Acoustically Lined Cylindrically Curved Ducts Without Fluid Flow. J. Sound Vibr., vol. 66, no. 2, Sept. 22, pp. 165–179.

Ko, S.H.; and Ho, L.T. (1977) Sound Attenuation in Acoustically Lined Curved Ducts in the Absence of Fluid Flow. J. Sound Vibr., vol. 53, no. 2, July 22, pp. 189–201.

Krasnushkin, P.E. (1945) Waves in Curved Tubes. NASA TT-20574, 1989 (Translation from Uch. Zap. MGU, No. 75, Book 2, Part 2, pp. 9–27).

Lippert, W.K.R. (1954) The Measurement of Sound Reflection and Transmission at Right-Angled Bends in Rectangular Tubes. Acustica, vol. 4, no. 2, pp. 313–319.

Luxton, R.E. (1968) A Low Pressure Loss Attenuating Bend for Airflow Ducts. The 6th International Congress on Acoustics, Y. Kohasi, ed., Elsevier, pp. 161–164.

Myers, M.K.; and Mungur, P. (1976) Sound Propagation in Curved Ducts. Aeroacoustics: Fan Noise and Control; Duct Acoustics: Rotor Noise, I.R. Schwartz et al., eds., Progress in Astronautics and Aeronautics, AIAA, vol. 44, pp. 347–362.

NBS (1964) *Handbook of Mathematical Functions*, National Bureau of Standards Applied Mathematics Series No. 55, U.S. Department of Commerce, pp. 435–478.

Osborne, W.C. (1968) The Propagation of Sound Waves Round a Bend of Rectangular Section. National College for Heating, Ventilating, Refrigeration and Fan Engineering Technical Memo No. 5.

Osborne, W.C. (1974) Calculation of the Angular Propagation Constant for a Bend. J. Sound Vibr., vol. 37, no. 1, Nov. 8, pp. 65–77.

Osborne, W.C. (1975) Propagation of Sound in a Bend Joining Two Straight Waveguides of Rectangular Cross-Section. Ph.D. Thesis, Council for National Academic Awards.

Osborne, W.C. (1976) Higher Mode Propagation of Sound in Short Curved Bends of Rectangular Cross-Section. J. Sound Vibr., vol. 45, no. 1, Mar. 8, pp. 39–52.

Prikhod'ko, V. Yu; and Tyutekin, V.V. (1982) Natural Modes of Curved Waveguides. Sov. Phys. Acoust., vol. 28, no. 5, Sept.–Oct., p. 422.

Rayleigh, J.W.S. (1945) *Theory of Sound*, Vols. I and II, 2nd edition, Dover Publications. (First edition, 1878.)

Rice, E.J. (1975) Spinning Mode Sound Propagation in Ducts With Acoustic Treatment. NASA TN D-7913.

Rice, S.O. (1948) Reflections From Circular Bends in Rectangular Waveguides—Matrix Theory. Bell System. Tech. J., vol. 27, no. 2, Apr., pp. 304–349.

Rostafinski, W. (1970) Propagation of Long Waves in Curved Ducts. Sc.D. Thesis, School of Engineering, Catholic University of Louvain, Belgium (NASA TM X-67184).

Rostafinski, W. (1972) On Propagation of Long Waves in Curved Ducts. J. Acoust. Soc. Am., vol. 52, no. 5, pt. 2, pp. 1411–1420.

Rostafinski, W. (1974a) Analysis of Propagation of Waves of Acoustic Frequencies in Curved Ducts. J. Acoust. Soc. Am., vol. 56, no. 1, July, pp. 11–15.

Rostafinski, W. (1974b) Transmission of Wave Energy in Curved Ducts. J. Acoust. Soc. Am., vol. 56, no. 3, Sept. pp. 1005–1007.

Rostafinski, W. (1974c) Zeros of Certain Cross Products of Bessel Functions of Fractional Order. NASA Tech Brief B74-10012. (See slso, NASA TM X-2698.)

Rostafinski, W. (1976) Acoustic Systems Containing Curved Duct Sections. J. Acoust. Soc. Am., vol. 60, no. 1, July, pp. 23-28.

Rostafinski, W. (1982) Propagation of Long Waves in Acoustically Treated, Curved Ducts. J. Acoust. Soc. Am., vol. 71, no. 1, Jan. pp. 36-41.

Tam, C.K.W. (1976) A Study of Sound Transmission in Curved Duct Bends by the Galerkin Method. J. Sound Vibr., vol. 45, no. 1, Mar. 8, pp. 91-104.

Ting, L.; and Miksis, M.J. (1983) Wave Propagation Through a Slender Curved Tube. J. Acoust. Soc. Am., vol. 74, no. 2, Aug., pp. 631-639.

NASA
National Aeronautics and
Space Administration

Report Documentation Page

1. Report No. NASA RP-1248	2. Government Accession No.	3. Recipient's Catalog No.
4. Title and Subtitle Monograph on Propagation of Sound Waves in Curved Ducts		5. Report Date January 1991
		6. Performing Organization Code
7. Author(s) Wojciech Rostafinski		8. Performing Organization Report No. E-5480
		10. Work Unit No. 505-69-61
9. Performing Organization Name and Address National Aeronautics and Space Administration Lewis Research Center Cleveland, Ohio 44135-3191		11. Contract or Grant No.
		13. Type of Report and Period Covered Reference Publication
12. Sponsoring Agency Name and Address National Aeronautics and Space Administration Washington, D.C. 20546-0001		14. Sponsoring Agency Code

15. Supplementary Notes

16. Abstract

After reviewing and evaluating the existing material on sound propagation in curved ducts without flow, it seems strange that, except for Lord Rayleigh in 1878, no book on acoustics has treated the case of wave motion in bends. This monograph reviews the available analytical and experimental material, nearly 30 papers published on this subject so far, and concisely summarizes what has been learned about the motion of sound in hard-wall and acoustically lined cylindrical bends.

17. Key Words (Suggested by Author(s)) Sound propagation Ducts Bends	18. Distribution Statement Unclassified – Unlimited Subject Category 71		
19. Security Classif. (of this report) Unclassified	20. Security Classif. (of this page) Unclassified	21. No. of pages 96	22. Price* A05

NASA FORM 1626 OCT 86 *For sale by the National Technical Information Service, Springfield, Virginia 22161

NASA-Langley, 1991